SpringerBriefs in Statistics

For further volumes:
http://www.springer.com/series/8921

Nicholas T. Longford

Statistical Decision Theory

 Springer

Nicholas T. Longford
SNTL and Department of Economics and Business
Universitat Pompeu Fabra
Barcelona
Spain

ISSN 2191-544X   ISSN 2191-5458   (electronic)
ISBN 978-3-642-40432-0   ISBN 978-3-642-40433-7   (eBook)
DOI 10.1007/978-3-642-40433-7
Springer Heidelberg New York Dordrecht London

Library of Congress Control Number: 2013948867

Printed on acid-free paper

Springer is part of Springer Science+Business Media (www.springer.com)

To Mr. A. Horger and his colleagues

Preface

Making decisions is at the heart of all human activity, mundane as well as adventurous, for both livelihood and fun. It is an abstract process in which we try to anticipate what may happen if we choose one course of action or another, and weigh carefully which one is safer, more enjoyable and profitable, or will bear greater benefits of some other kind. The definition of statistics assumed in this volume is that of making decisions in the presence of uncertainty and with limited resources. In this perspective, any meaningful analysis should closely look at the consequences of the possible outcomes and of the recommendations based on them.

Uncertainty means that we may get it wrong. For example, even a competently executed test of a hypothesis may conclude with evidence against the null when in fact the null is valid. We do not allow a 5 % chance when crossing a road known to be busy—it makes sense to look at and listen to the traffic, appreciating the consequences of getting this everyday manouevre wrong. In a statistical analysis, the consequences of the errors we commit may or may not be as lop-sided as in this example, but their balance is a relevant factor in the analysis.

To my regret, I have come to this view only recently, prompted by my experiences in consulting in which some clients have dismissed the standard format of an analysis that concludes with a hypothesis test, a confidence interval, or an estimate, because such a statement requires further nontrivial processing (interpretation) to decide what to do: how to alter business practices, production settings, strategic goals, medical treatment regimes, and the like. I got the drift and re-discovered the work of Morris DeGroot, Tom Ferguson, Dennis Lindley and Jim Berger, and this volume is a document of my conversion—not to a Bayesian, but to regarding a decision as the ultimate goal of an analysis.

The volume has an unashamed partisan character; I would like to convert the reader to the perspective I present, or at least to appreciate its merits, together with its practical feasibility and the potential appeal to a client.

Chapter 1 introduces the main ideas and terminology, discusses the established formats in which the results of statistical analyses are most frequently presented, and highlights their unsatisfactory features. Chapters 2 and 3 deal with two elementary problems, estimation of the mean and variance of a normal random sample, when the consequences of errors with one sign differ from those with the other sign. The problem of choosing one of the options (courses of action) is also addressed. Chapter 4 introduces the Bayesian paradigm as a vehicle for incorporating prior information about the parameters involved in a model. These priors are informative and are regarded as important as the data. Chapter 5 steps beyond the confines of the normal distribution and shows that decisions with samples from and statistics with some other distributions involve calculus not substantially more complex than in the normal case. Chapters 6 and 7 deal with two applications, classification and small-area estimation, in which, I believe, the absence of a transparent discussion of losses (utilities) has led the development astray, sometimes in the direction of irrelevance. Chapter 8 deals with study design. Its length and location in the volume do no justice to the importance of design in practice, but without having settled on the method of analysis its discussion would not be constructive.

All the evaluations described in this volume were made in R (R Development Core Team, 2009), with user-defined functions. Apart from working out all the examples, this also had the purpose of quality control of all the math expressions in the text. The entire library is available from www.sntl.co.uk/Decide.html. However, a reader may draw greater benefit from the volume by implementing some or all the algorithms himself or herself, exploring alternative graphical displays and assessing firsthand the programming effort required.

The original intention of the volume was to communicate the relevance, feasibility, and value of decision theory for the 'everyday' statistical problems to (statistical) professionals, but I paid attention during the revisions also to the potential of using the text in a graduate course in statistics. A good background in calculus and linear algebra is important and proficiency in R is a distinct advantage. With it, the text can be used in a graduate course in one academic quarter. The necessary background from other textbooks can be combined with this text in a semester. If a topic has to be dropped from the course, Chaps. 6 or 7 are better choices than Chap. 8.

I would like to thank Aleix Ruiz de Villa Robert for insightful comments on parts of the manuscript, to Omiros Papaspiliopoulos for highly informative discussions, and to Springer-Verlag for the original idea, assistance with preparation of the manuscript, arranging helpful reviews of a draft manuscript and smooth and timely production of the monograph. The project was partly supported by a grant from the Spanish Ministry of Science and Innovation.

Barcelona, April 2013

Contents

Chapter 1
Introduction

Most statisticians are accustomed to concluding the formal part of their analysis by one or several estimates, confidence intervals or by the verdict of a hypothesis test, formulated as either failure to reject or rejection of the hypothesis. We refer to them as *inferential statements*. In some simple settings, there is a clear computational protocol for producing such a statement, which can be traced to the general theory for the underlying principles and rationale. The principles are not complete, and have to be supplemented by some widely adopted conventions that have a questionable foundation, such as the choice of 5 % for the level of significance (the size) of the test. (There is no profound reason for not using 4.5 % instead.) There may be arguments about the appropriateness of the assumptions that correspond to the adopted setting, such as normality, independence and linearity in an ordinary regression, and diagnostic procedures offer some ways of arbitrating about them.

Throughout this book, we assume that an analysis is conducted on the request of a client who has a particular agenda, looking for the answer to a particular question that pertains to the analysed data. The analyst's formal inferential statement is followed by its *interpretation*, a discussion of how the result relates to the client's problem, that is, a translation of the result to the language and context of the client. In this step, the traditional textbook offers plenty of examples but few principles, arguing, in essence, that every problem is different, and that practical experience is the best instructor.

If the communication between the analyst and the client is not sufficiently close, the interpretation, or part of it, may be left to the client. The thesis put forward in this chapter, and developed in the following chapters, is that this 'interpretation' is a nontrivial activity that requires evaluations that are firmly in the remit of a statistical analyst. Moreover, it can be connected to basic scientific principles, and can thus be formalised. This approach is universal, applicable to a wide range of problems that are generally regarded as statistical, but its application is computationally challenging in some complex settings. However, sparing computers is a dubious scientific strategy.

N. T. Longford, *Statistical Decision Theory*,
SpringerBriefs in Statistics, DOI: 10.1007/978-3-642-40433-7_1,
© The Author(s) 2013

1.1 The Role of Statistics

Our starting point is a definition of statistics as a profession or scientific field. We define its role as

making purposeful decisions in the presence of uncertainty with limited resources.

The key terms in this definition, purposeful decision, uncertainty and limited resources, are discussed next.

A purposeful decision is what the client wants to make in the pursuit of a professional goal, and why he or she approaches a statistician, or engages in a statistical activity. The underlying purpose, or goal, may be to save resources in a manufacturing process, confirm the quality of a production process or an innovation, improve the services provided or understand a natural or social phenomenon. For most part, we reduce our attention to the task of choosing between two courses of action (options), A and B. These actions are exclusive—one of them has to be taken, and they are complementary — it is impossible to take both of them. Note that taking 'no action', such as dismissing a suspect from a police station without any charge, is itself a course of action; its complement is charging the suspect (and following it up by the standard police procedures). Dealing with any finite number of options, A, B, C, ..., Z, can be organised into a set of choices between two options. First we decide between A and the union of B, C, ..., Z (B–Z, for short); if we choose the union, we next decide between B and C–Z, and so on, until we choose a singular option, such as P, or end up having to choose between Y and Z.

Uncertainty refers to incompleteness of the available data. For example, an experiment can be conducted on only so many units, because the conduct consumes (client's) resources. The resources include materials, funding, time, but also the goodwill of the (human) subjects involved in the experiment. We can (and should) consider also ethical costs; we are bound by codes of good practice and laws that prescribe minimum exposure of human subjects, animals and the environment in general to any potential harm.

Complete data refers to having information based on which the client could choose among the contemplated courses of action without any difficulty. For example, if we could establish how a particular compound (a drug) suppresses certain symptoms of a disease in every patient who is suffering (or will in the future suffer) from a particular disease, the developer's decision would be simple and instant. However, the drug can be administered experimentally only to a small or moderate number of subjects, and only one such study is permitted by the regulatory authority. As a result, we have to work with incomplete information. Another source of incompleteness is that the observations made in the process of data collection are not precise (clinical); measurements are subject to rounding, imperfect instrumentation and they involve subjects' verbal responses (including opinions) that require the data collector's interpretation (coding). They are affected by momentary influences that include poor disposition, fatigue brought on by the interview and everyday distractions.

As an aside, we remark that statistics is not solely about data analysis. The client may (and often should) consult the statistical analyst in the planning stage, before expending resources on the conduct of a study and data collection. *Design* of a study is a key element of statistics. It can be regarded as an exploration of how reliable the results of a study will be depending on how the study is organised. That is, the statistician's remit includes advice on how to expend the available resources on a study (to collect data) in a manner that best serves the client's interests.

Apart from the expertise generally associated with statistics, serving these interests entails the ability to conduct a dialogue with the client in which the relevant details are elicited. The details include background to the (planned or already concluded) study, information about related studies and the subject matter in general that would help the analyst identify an appropriate way (method) of solving the problem and, foremost, the client's purpose (perspective). The premise of the dialogue is that the analyst would identify the best course of action at a particular juncture in the client's business, had it not been for incompleteness of the available information.

Having access only to incomplete information, the analyst's answer may be incorrect. That is, if the complete information were available, the correct answer (the appropriate course of action) would be identified, whereas with the available information the analyst can only do the best that can be done. But the client has a key input into what the qualifier 'best' means in this context, and the basis of this qualification is an assessment of the losses (damage, harm, additional expense, and the like), if instead of the optimal another course of action were taken. Another client may assess the consequences of these errors differently—the analysis is tailored to the particular client's interests, perspectives and purpose. In this respect, the statistical analysis is subjective.

A disclaimer is in order. The error mentioned earlier is not a fault of the analyst, and does not imply incorrectness of his or her analysis. It is an inevitable consequence of the incompleteness of the information available for the analysis. However, the analyst should ensure that, in some well defined way, the consequences of such errors are minimised. This 'way', or criterion, has to be agreed with the client. A default criterion is a simple solution, but it may be found wanting later. The criterion is applied by an interplay of the design (what kind of data to collect, and how) and analysis (how to process the collected data). Attention to one aspect is not sufficient, although one cannot do much about the design once the study (data collection) has been completed.

Incompleteness of information should never serve as the analyst's excuse, and should not be regarded as one side of a dichotomy, completeness being the other. Incompleteness is an implaccable adversary. We rarely have the capacity to eradicate even some of its minor aspects without some sacrifice that impedes our ability to combat some other aspects. A better strategy is to carefully target the available resources on the aspects that are most detrimental to our client's interests.

Often there is ample scope for identifying sources of information that would supplement the focal source. Such information may be in a form other than a dataset, and its incorporation in the analysis may be challenging, but the related 'detective' work should be motivated by the prospect of higher quality of the analyst's output, and better service to the client.

In summary, we seek in statistics to do the best that can be done with the available resources. The 'best' is defined by a criterion tailored to the problem at hand, reflecting the perspective and priorities of the client, and 'resources' are interpreted broadly as information, including data, together with our limited capacity to collect (new) data in a purposeful way.

Design logically precedes analysis, but we will treat these two subjects in the reverse order; design is easier to motivate when the issue of how to analyse the data has been settled. For developing methods of analysis, we will assume certain standard (default) designs.

The next section introduces the elementary terms used in the book. This is necessary both for completeness of the text and because some of the terms are used inconsistently in the literature.

1.2 Preliminaries

We will work with two primitive terms: population and variable. A *population* is defined as a collection of units. Such a definition is proper only when there is absolutely no ambiguity about any entity as to whether it is a member of the population (one of the units) or not. A *variable* is defined in a population by a rule that assigns to each member a particular value. The value does not have to be numeric. The support of a variable is defined as the set of all values that the variable attains in the population. With such values we associate operations that are well defined for them. For example, comparison is an operation on ordered pairs of values. For any pair x and y, one of the three relations holds: $x < y$, $x = y$ or $x > y$. They are complementary and exclusive. Further, $x < y$ implies that $y > x$; $x < y$ and $y < z$ implies that $x < z$; and $x = y$ and $y = z$ implies that $x = z$.

For a variable (defined in a population), we can define various summaries. A simple example for a variable with numeric values is its (or their) mean. We assume that its evaluation would require next to no effort if the value were available for every member of the population; that is, if in the narrow context of this population and the variable we had complete information.

When the values of the variable are established (recorded, measured, elicited, or made available by some other means) for only some members of the population, we say that the variable is observed on a sample of units. We can treat this sample as a population, and define for it a summary, such as the mean. We refer to it as the sample mean. More generally, we define population and sample summaries of a variable. In our example, the population mean has the sample mean as its sample counterpart.

In a particular statistical perspective, called frequentist, a population summary is regarded as a fixed quantity, irrespective of whether its value is known or not. This reflects the state of affairs in which its value is unaffected by our efforts to establish it. In contrast, a sample mean depends on the units (subjects) that have been (or will be) included in the sample. If the process of selecting (identifying) a sample entails some randomness (happenstance or uncertainty), then the sample quantity also entails some randomness—a different sample may equally well have been selected, and it would yield a different value of the quantity.

In the frequentist perspective, we consider *replications* (independent repeats) of the process by which the data is generated and then processed to yield an inferential statement. The quality of such a statement is assessed by a reference to replications. The ideal, usually not attainable, is that every replicate statement is correct, that is, it either does not differ from the population quantity that is the target of the inference or is not in a contradiction with it. This ideal corresponds to the absence or irrelevance of any uncertainty, coupled with an appropriate way of processing of the data (the sample values). When there is some uncertainty the replicate statements differ (vary), and we conventionally assess them by how frequently they are correct or how distant they tend to be from the correct statement (the target).

When the conduct of a study is expensive, as is often the case, we can afford only one replication of the study. The resulting dataset is referred to as the realised version of the study, and the statement based on it as the realised statement. Thus, the assessment of an inferential statement is abstract, relying on the design of the study and its setting to infer what datasets and statements would be obtained in *hypothetical* replications of the study. Here 'study' means a protocol for the conduct of a study; the other meaning of 'study' is its realised version, the one-time implementation of the study protocol.

The role of a theory is to identify statements about population quantities, that is, protocols for the production of such statements, that would be favourably assessed by the criterion agreed with the client. Of course, such statements are sample quantities, subject to uncertainty. A particular challenge for the theory is that the correct statement is not known.

An alternative to such a theory is offerred by the computer. We construct a computer version of the data-generating process that can be inexpensively executed many times, generating replicate datasets, and formulate the inferential statement based on each replicate. These replicate statements are then assessed as originally intended — how close they are to the target on average or how frequently they match it. This process of generating and evaluating a set of replicate datasets is called *simulation*.

Neither a theory nor a computer simulation are perfect. The theory is precise and reliable only in some stylised settings and cannot always be extended (extrapolated) to the setting of a particular study. The computer may be more flexible, but the data-generating process is rarely known completely. This problem is addressed by simulating a range of plausible scenarios. The assessments based on them indicate the plausible quality of the studied inferential statements. This approach, called *sensitivity* analysis, entails an open-ended process, limited by the capacity of our computer, but also by our programming skills—implementing the procedures so that they can easily be adapted to a range of scenarios and executed efficiently, using as little processing time as possible.

A sample has two versions. Prior to the conduct of the study, which starts with identifying the members of the population that will become the subjects (units, or elements) in the sample, the sample is a random entity. After their identification (the draw of the sample), it is fixed. Any sample quantity therefore also has two versions.

While the sample is random, the sample quantity is also random—there is uncertainty about its value. When the sample is realised and the data collected, the quantity is fixed. An inferential statement also has two versions: it is a protocol for evaluation, described by mathematical equations and implemented (or implementable) on a computer, and the actual statement, such as a number, an interval, or the selection from a finite set of mutually exclusive and exhaustive options.

In the following sections we discuss the most common formats of inferential statements.

1.3 Estimation

A typical study is conducted to learn (more) about one or several population quantities; we refer to such quantities as *targets* (of the study). We focus on a single target, and assume it to be numeric. One obvious solution is the sample counterpart of this quantity. It is an example of an estimator and estimate. Estimator is the version associated with a random sample and estimate with the realised sample. The estimator is a formula or a computer programme that implements it, and the estimate is a number, such as 7.32, obtained by applying the estimator on the realised sample.

The purpose of estimation is to obtain a sample quantity that is as close as possible to the target, a population quantity. The target is denoted by θ and an estimator by $\hat{\theta}$. When we consider several estimators of the same target (after all, there is meant to be a contest for estimating θ), we use subscripts or superscripts with $\hat{\theta}$ and we use also the notation $\tilde{\theta}$.

The estimation error is defined as $\hat{\theta} - \theta$. When we do not care about the sign of the error we can reduce our attention to the absolute (estimation) error $|\hat{\theta} - \theta|$. With it, zero is the ideal; small is good and large is bad. This applies equally to any increasing transformation g for which $g(0) = 0$. The squared error $(\hat{\theta} - \theta)^2$ is such a transformation. Note that with a transformation we have to adjust our scale for the meaning of 'small' and 'large'.

The mean squared error of an estimator $\hat{\theta}$ for the target θ is defined as the average of the squared errors $(\hat{\theta} - \theta)^2$ in a long sequence of replications of $\hat{\theta}$. We write

$$\mathrm{MSE}\left(\hat{\theta}; \theta\right) = \mathrm{E}\left\{\left(\hat{\theta} - \theta\right)^2\right\}.$$

An advantage of this assessment of an estimator is that it can be expressed in terms of two familiar characteristics of the estimator: its sampling variance and bias:

$$\mathrm{MSE}\left(\hat{\theta}; \theta\right) = \mathrm{var}\left(\hat{\theta}\right) + \left\{\mathrm{B}\left(\hat{\theta}; \theta\right)\right\}^2,$$

where var denotes the sampling variance,

$$\mathrm{var}\left(\hat{\theta}\right) \;=\; \mathrm{E}\left[\left\{\hat{\theta} - \mathrm{E}\left(\hat{\theta}\right)\right\}^{2}\right],$$

and *B* is the bias,

$$\mathrm{B}\left(\hat{\theta}; \theta\right) \;=\; \mathrm{E}\left(\hat{\theta}\right) - \theta.$$

The bias represents a systematic component of the error and the sampling variance the uncertainty (variation or dispersion) of the estimator. Note that the arguments of MSE and bias include the target θ. An estimator could be used for more than one target, and the bias (and MSE) for one may differ from the bias (and MSE) for the other. In contrast, the variance does not depend on the target. The bias and variance need not be defined, so MSE is not always suitable for assessing the quality of an estimator or for comparing the quality of alternative estimators of the same target. However, we need not be concerned with this for the time being; we lose next to no generality by assuming that both bias and variance are finite, or by dismissing any estimator with undefined bias or variance as outright unsuitable. The MSE is a very popular criterion, but we will develop the view that this is partly due to the neglect of some important aspects of the client's perspective.

An estimator is called *efficient* for a target if its MSE is well defined and no other estimator has a smaller MSE. Of two estimators of the same target, one is said to be more efficient than the other if its MSE is smaller. We emphasise that *we* chose to assess the quality of an estimator by MSE. This choice is a convention, supported by convenience—its relatively easy evaluation and motivation. It entails the assumption that positive estimation errors, $\hat{\theta} > \theta$, have as serious consequences for the client as do negative errors, $\hat{\theta} < \theta$, of the same magnitude $|\hat{\theta} - \theta|$. Our client may find this unreasonable, as he may find the assumption that twice as large an error is regarded as four times more serious. We should not attempt to convince him that adhering to the convention is advantageous if such a treatment of the estimation errors is not in accord with his purpose and perspective. It may serve some of the analyst's interests, but these are narrow and secondary. On the contrary, we should stimulate the client's interest in this issue, because it is essential for taking into account his perspective in the analysis, and thus serve his interests better.

The mean absolute error is defined as

$$\mathrm{MAE}\left(\hat{\theta}; \theta\right) \;=\; \mathrm{E}\left(\left|\hat{\theta} - \theta\right|\right).$$

With MAE, twice as large an absolute error is twice as serious. We refer to MSE and MAE as criteria (for the quality of estimators). They can also be interpreted as expected losses (expectations of loss functions); simply, we interpret the relevant transformation of $|\hat{\theta} - \theta|$ as the loss. It is much more difficult to work with MAE, but that should not discourage us from adopting it if it reflects the client's perspective well, better than an analytically more convenient alternative.

Sensitivity analysis, in effect, producing the statement that is best with MSE and one that is best with MAE, and others that are superior with other assessments of the estimator (criteria for estimation) may compensate for the analyst's inability to agree with the client on the criterion. If the same (or similar) statements are obtained with all these criteria, the choice among them is not important. Otherwise, elicitation from the client is essential, or our conclusion has to be qualified by (conditioned on) the criterion.

1.4 Assessing an Estimator

The root-MSE is defined as the square root of the MSE, $\sqrt{\text{MSE}(\hat{\theta}; \theta)}$, of estimator $\hat{\theta}$ for target θ. The standard error is defined as the square root of the sampling variance. For estimators with no bias, $B(\hat{\theta}; \theta) = 0$, MSE is equal to the sampling variance; then the root-MSE and the standard error coincide.

Two estimators of the same target can be compared by their root-MSE; the estimator with smaller root-MSE is preferred. However, MSE may, and often does, depend on the value of the target θ or of another population summary ξ. Then one estimator may be more efficient than the other for some values of θ and ξ, but not for others, so there is no straightforward way of comparing the two estimators. Comparison based on $\text{MSE}(\hat{\theta}; \hat{\theta})$, substituting the estimate $\hat{\theta}$ also for the target, is usually flawed. For two estimators, $\hat{\theta}_1$ and $\hat{\theta}_2$, comparing $\text{MSE}(\hat{\theta}_1; \hat{\theta}_1)$ with $\text{MSE}(\hat{\theta}_2; \hat{\theta}_2)$ is not of like with like. The comparisons of $\text{MSE}(\hat{\theta}_1; \hat{\theta}_2)$ with $\text{MSE}(\hat{\theta}_2; \hat{\theta}_2)$ and of $\text{MSE}(\hat{\theta}_1; \hat{\theta}_1)$ with $\text{MSE}(\hat{\theta}_2; \hat{\theta}_1)$ may be in discord, but even if they are in agreement they may be in discord with the comparison based on the (unknown) value of θ. The absence of a clear-cut answer is not a deficiency; it is an integral part of the overall uncertainty that in general should be neither ignored nor resolved by improvisation or by adopting a superficially agreeable convention.

An estimator of a target is said to be uniformly more efficient than another estimator of the same target, if it is more efficient for every value of θ and other population quantities. An estimator is said to be *admissible* for a target if there is no other estimator that is uniformly more efficient for the same target. There is no virtue in using an inadmissible estimator $\hat{\theta}$, except when we cannot find an estimator uniformly more efficient than $\hat{\theta}$, despite knowing that such an estimator exists. The MSE can be evaluated for a range of plausible values of θ and regarded (and studied) as a function of θ and other population quantities.

The MSE of a typical estimator is not known and has to be estimated from a single realised sample. For such an estimator we can adopt the same criteria (forms of assessment) as for other estimators, although this can very quickly become both a mouthful and a never-ending task, because the estimator itself has a MSE, which itself may require estimation. Therefore, we might adopt the convention that we will care only about the bias of the estimator of a MSE, and regard its unbiased estimation as the ideal.

By estimating MSE (root-MSE or MAE), we are assessing the quality of an estimator, the product of our labour. Thus, we are our own clients. In such a position, we have to resist any temptation to present an estimator in a better light than it deserves. We propose an approach motivated by the attitude projected by any self-respecting supermarket or another retail establishment. We regard overestimation (positive bias) of MSE by $\Delta = E(\widehat{MSE}) - MSE > 0$ as much less serious than underestimation (negative bias) by $-\Delta$. A parallel can be drawn with buying an item in a supermarket. On its packaging we read that it should be used by a certain date, and its good quality until then, with the appropriate disclaimers, is guaranteed. If we consume the item after this date and its quality has not deteriorated, we do not owe anything to the supermarket. However, if the quality has deteriorated before the date of the warranty, we can return the item and will receive a refund or replacement (or both), with apologies. Arguably, overstating the quality (durability) is treated rather harshly in this context. In contrast, understating the quality has no consequences, except perhaps that the producer (and the supermarket) could have made a stronger claim, both on the packaging and in the marketing arena, and maybe could have charged a higher price for a product with a quality rightly claimed to be higher.

Similarly, an analyst loses face if it later turns out that he underestimated the MSE of an estimator. Unbiased estimation of MSE is universally adopted as the standard, but the 'supermarket' viewpoint suggests that we should overestimate it. More precisely, we should be concerned not about the bias, but about the sign of the estimation error $\widehat{MSE} - MSE$ and, ideally, avoid negative estimation error altogether. That is perhaps a tall order, but adhering to an objectionable convention in its stead is equally problematic. We revisit this issue in Chap. 3.

1.5 Confidence Intervals

A confidence interval for a population quantity θ is defined as an interval $(\hat{\theta}_L, \hat{\theta}_U)$ in which both limits are sample quantities. The intent of a confidence interval is to indicate where the target θ is likely to be. A confidence interval is qualified by a probability, or percentage. A 95 % confidence interval is constructed with the intent that in a large number of hypothetical replications θ, a fixed quantity, would be contained in the interval $(\hat{\theta}_L, \hat{\theta}_U)$, with random bounds, in at least 95 % of the replicates;

$$P\left\{\theta \in \left(\hat{\theta}_L, \hat{\theta}_U\right)\right\} \geq 0.95.$$

A confidence interval is called *proper*, or is said to have the claimed coverage, if this condition is satisfied. A typical reason for why a confidence interval is not proper is that the assumptions adopted in its construction are not satisfied. A confidence interval may be proper for some values of θ and other population quantities, but not for others. The confidence interval $(-\infty, +\infty)$ is proper but not useful, because it conveys a vacuous statement.

A confidence interval is called one-sided if $\hat{\theta}_L = -\infty$ or $\hat{\theta}_U = +\infty$. Otherwise, when its width $\hat{\theta}_U - \hat{\theta}_L$ is finite, it is called two-sided. Two one-sided confidence intervals for θ can be meaningfully compared only when either $\hat{\theta}_L = -\infty$ or $\hat{\theta}_U = +\infty$ for both of them; a confidence interval that is contained in the other is preferred, so long as it is proper. Among alternative two-sided confidence intervals for a target θ, the one that is narrower (has shorter length) is preferred, so long as it is proper. With a higher standard, a confidence interval is said to be better than another if it is proper and is contained in the latter. This standard defines a partial ordering—some pairs of confidence intervals cannot be compared. A confidence interval is called admissible if it is proper and none of its sub-intervals (except for itself) is proper.

A two-sided confidence interval for the mean of a population is often constructed as $(\hat{\mu} - c\hat{\sigma}, \hat{\mu} + c\hat{\sigma})$, where $\hat{\mu}$ is an unbiased estimator of the mean μ, $\hat{\sigma}^2$ an unbiased estimator of the sampling variance of $\hat{\mu}$, and c a suitable constant set so as to ensure that the interval has the proper coverage. Thus, a confidence interval can be constructed from an unbiased estimator and an unbiased estimator of its sampling variance.

Confidence intervals represent the uncertainty about the target in a simplistic way. The price for simplicity is incorrectness. A confidence interval $(\hat{\theta}_L, \hat{\theta}_U)$ is usually interpreted, or the client acts upon its receipt, by ruling out the possibility that θ might lie outside this interval. With a proper confidence interval, the probability that this statement is correct is at least 0.95. But the consequences of θ lying outside the interval, admittedly not a very likely event, may be disastrous. This is a distinct weakness of the confidence interval in general. An estimate is associated with similar incorrectness, if the analyst or his client treats it, and acts upon it, as if it were the target.

1.6 Hypothesis Testing

The intent of hypothesis testing is to resolve the issue as to whether the value of a population quantity θ is in one set, H_0, called the hypothesis, or another, H_1, called the alternative. The two sets are exclusive, although they need not be separated. For example, H_0 may be that $\theta = 0$, and H_1 that $\theta > 0$. A hypothesis or an alternative that contains a single value is called simple. We deal first with a test of a simple hypothesis against a simple alternative.

The roles of the hypothesis and alternative are not symmetric. In essence, the hypothesis is regarded as the default, and is rejected only when there is sufficient (data-based) evidence against it. The possible outcomes of testing a hypothesis are its rejection and failure to reject it. The conduit for the evidence is a sample quantity called the test statistic. Its random version is selected by the analyst. It is then evaluated on the realised sample. Denote the statistic by t and its value by $t(\mathbf{y})$, where $\mathbf{y} = (y_1, \ldots, y_n)$ is the vector of the values of the relevant variable on the sample.

Next we study the distribution of t assuming that the hypothesis is valid. If the realised value $t(\mathbf{y})$ is exceptional with respect to this distribution, then we regard it as evidence against the hypothesis. We have several options for defining the term 'exceptional'. First we specify the size of the test, a small probability, denoted by

α. It is the intended probability of rejecting the hypothesis when the hypothesis is valid. The convention is to set α to 0.05, although 0.01 and 0.10 are also used, though rarely.

In the symmetric test of a hypothesis, we identify the critical values c_L and c_U such that $P_0(t < c_L) = P_0(t > c_U) = \alpha/2$; the subscript 0 indicates evaluation with the assumption of the hypothesis. With a slight abuse of notation, we could write more explicitly $P(t < c_L \mid H_0)$. The abuse would arise because H_0 is not an event that is associated with a (positive) probability. The values in the interval (c_L, c_U) are regarded as 'not unexpected', and those outside as exceptional, *assuming* H_0. We reject the hypothesis, if the value $t(\mathbf{y})$ is exceptional, when $t(\mathbf{y}) < c_L$ or $t(\mathbf{y}) > c_U$.

Note that when we fail to reject the hypothesis, $t(\mathbf{y})$ is not necessarily an exceptional value for the alternative. That is, let $d_L < d_U$ be such that $P_1(t < d_L) = P_1(t > d_U) = \alpha/2$, where the subscript 1 indicates that the alternative is assumed. Then $t(\mathbf{y})$ may lie inside the interval (d_L, d_U). But $t(\mathbf{y})$ may be an exceptional value for both the hypothesis and the alternative. In brief, the roles of the hypothesis and the alternative cannot be interchanged.

The set of values for which the hypothesis is rejected is called the critical region of the test. In the setting introduced, this set is $\mathcal{C} = (-\infty, c_L) \cup (c_U, +\infty)$. A test is called one-sided if the critical region is an interval, $(-\infty, c_L)$ or $(c_U, +\infty)$, but not their union. In principle, any subset of the support of t can form the critical region; it does not have to be an interval or a union of two intervals, although examples when it is neither are difficult to construct in commonly encountered settings.

The power of a test is defined as the probability, evaluated under the alternative, that the hypothesis is (correctly) rejected; it is denoted by β. That is, $\beta = P_1(t \in \mathcal{C})$. A desirable property of a test with a given size α is that β is large. Thus, a test is called more powerful than another test of the same hypothesis against the same alternative, and with the same size α, if its power is greater. The Neyman-Pearson lemma (Neyman and Pearson 1933) states that under certain assumptions the *likelihood ratio test* is most powerful.

A test is said to have a complex hypothesis if the hypothesis comprises more than one value. The critical region has to be constructed in such a way that its probability does not exceed the size of the test for any value in the hypothesis. Denote the set of values of θ in the hypothesis and alternative by Ω_0 and Ω_1, respectively. The same procedure for testing the hypothesis is followed as for a simple alternative, but the condition for the size of the test, $P_\theta(t \in \mathcal{C}) \leq \alpha$, has to be satisfied for every $\theta \in \Omega_0$.

With a complex alternative, the power of the test is a function defined on the set Ω_1, and a comparison of two tests (with the same H_0 and the same H_1) is no longer straightforward. One test of size α is said to be uniformly more powerful than another of the same size, if the power function of the first test is greater than for the second for every value in the alternative Ω_1.

We define a sample quantity t with the intent to use it as the test statistic in a hypothesis test of prescribed size α. A hypothesis test is called proper if $P_\theta(t \in \mathcal{C}) \leq \alpha$ for every $\theta \in \Omega_0$. This probability may depend also on some other (unknown) population quantities. We may fail to ensure that a hypothesis test is proper for a myriad of reasons, including those listed in connection with improper confidence intervals.

They can be summarised as failure of some of the assumptions. A hypothesis test is called unbiased if $P_\theta(t \in C) \geq \alpha$ for every $\theta \in \Omega_1$.

For a simple hypothesis, we can construct a hypothesis test for θ from a confidence interval for θ. Simply, we reject the hypothesis if θ_0 lies outside the confidence interval. Thus, the pair of an (unbiased) estimator and estimated standard error can be re-packaged into the format of a confidence interval and a hypothesis test by simple operations. Therefore, we need confidence intervals and hypothesis tests only for presenting the results in a format that would be well accepted by the client. But this would still leave the client with the non-trivial task of translating the inferential statement into a plan of action—a decision. Some hypothesis tests are constructed without any reference to an estimator or a confidence interval, but nevertheless remain an unfinished product for the client.

Considering the format of its statement, a hypothesis test may at first look well suited for the problem of deciding between two courses of action, one corresponding to the hypothesis and the other to the alternative. However, the asymmetric roles of the hypothesis and the alternative are inappropriate for the task of decision making. There is no obvious way of informing the process of hypothesis testing about the consequences of the two kinds of bad decisions that can be made. The error of rejecting the hypothesis when it is in fact valid is called the type I error. For a proper test, its probability (assuming H_0) does not exceed the size of the test, α. For a complex hypothesis, this probability is a function of θ in Ω_0. The error of not rejecting the hypothesis when the alternative is valid is called the type II error. Its probability is equal to the complement of the power (function), $1 - \beta$, or $1 - \beta(\theta)$.

The incompatibility of hypothesis testing with decision making is illustrated on the following example. Suppose the hypothesis is that a particular population quantity θ is equal to zero, and the alternative is that $\theta \neq 0$. Suppose a suitable test statistic is found, and its distribution under H_0 is unimodal and symmetric. A symmetric critical region is then reasonable. However, somebody with a stake in the outcome of the test, who does not regard zero as a value with any exceptional status (e.g., different from 0.00021 or 12.0035) would not select H_0 because it is a bet against vast odds of a continuum of alternative values. Arguably, the value of θ may be quite small, but testing that hypothesis corresponds to a different procedure, for which we should first specify the meaning of the term 'small'. Even with this adaptation, we have the disconcerting asymmetry that H_1 is adopted when there is sufficient evidence for it, whereas H_0 is adopted not when it is supported by sufficient evidence, but merely when we lack sufficient evidence against it. The default status of the hypothesis is highly problematic.

1.7 Loss Function

We consider the problem of choosing between two courses of action, A and B. They are complementary and exhaustive. We quantify the consequences of the two kinds of error, choosing A when B is appropriate and choosing B when A is appropriate,

Table 1.1 Example of a loss function

Statement	Action	
	A	B
0 (A)	0	1
1 (B)	10	0

by a loss function. We associate every combination of a statement (based on incomplete information) and the optimal course of action (that could be established if the complete information were available) with a nonnegative quantity.

When there are only two options for the statement, this function of two arguments can be presented in a two-way table. An example is given in Table 1.1. There are two possible statements, 0 and 1, and two actions, A and B. Statement 0 is appropriate (correct) for action A and statement 1 for B. For these two combinations, the function vanishes, indicating that no loss is incurred. If statement 0 is made, but action B is appropriate, the loss is one unit. If statement 1 is made, but action A is appropriate, the loss is ten units. We are ten times as averse to inappropriately choosing B (following statement 1) as we are to inappropriately choosing A (following statement 0).

Declaring the loss for the correct decision as zero is a convention. The essence of the setting would not be altered if we added the same positive constant to every entry of the table. This constant can be interpreted as the running costs of the client's enterprise, and they are taken for granted. At issue is only the additional expense associated with the incorrect decision. We can also multiply each entry of the table by a positive constant. This corresponds to the change of currency in which the loss is quantified.

Greater loss is worse—that is implied by the term 'loss'. A well defined loss also has the property of additivity. That is, the loss of 10 units in the combination (1, A) in Table 1.1 is for all purposes equivalent to the harm or damage of ten instances of the unit loss in the combination (0, B). In general, a loss of a units in one instance followed by b units in another is equivalent to the loss of $a + b$ units in a single instance. Without additivity, averaging its values in a set of hypothetical replications would make no sense. The version of this average with infinitely many replications is the expected loss, $E(L)$. The loss function L has two arguments, the decision (the selected action), denoted by d, and the value of the parameter (vector) θ which determines which action is appropriate; $L(d, \theta)$. When the loss depends on θ only through the appropriate action, D, as it does in Table 1.1, we write $L(d, D)$.

Suppose the conditional probability of statement 0, given that A is the appropriate action, is $P(0 \,|\, A) = 0.90$, and $P(1 \,|\, B) = 0.75$. Then the probabilities of the two kinds of error are $P(1 \,|\, A)\,P(A)$ and $P(0 \,|\, B)\,P(B)$, and the expected loss is

$$L(1, A)\,P(1 \,|\, A)\,P(A) + L(0, B)\,P(0 \,|\, B)\,P(B) = 10 \times 0.10P(A) + 0.25P(B).$$

The marginal probabilities $P(A)$ and $P(B)$ are essential for evaluating the expected loss. If we could rule out action A, $P(A) = 0$, it would be wise to choose B, because no loss would be incurred when $P(B) = 1$, when B is bound to happen.

The loss function may depend on some other quantities. For example, A may be the appropriate course of action when a population quantity θ is smaller than a given threshold T, and B when $\theta \geq T$. Then a natural way to proceed might be to estimate θ, by $\hat{\theta}$, and pursue action A if $\hat{\theta} < T$ and action B otherwise. When $\hat{\theta}$ and θ imply different actions, that is, when T is located between θ and $\hat{\theta}$ ($\hat{\theta} < T < \theta$ or $\theta < T < \hat{\theta}$), the loss function depends, in general, on both θ and $\hat{\theta}$. So, L may be an increasing function of $|\hat{\theta} - \theta|$, or of the difference on another scale:

$$L = L_A\left(\hat{\theta}; \theta\right),$$

when $\hat{\theta} < T$ but $\theta > T$, and $L = L_B(\hat{\theta}; \theta)$, another increasing function of $|\hat{\theta} - \theta|$, when $\hat{\theta} > T$ but $\theta < T$. That is, we want to establish only the sign of $\theta - T$, but when we get it wrong, the loss depends on both the magnitude *and* the sign of the error $\hat{\theta} - \theta$.

1.8 Problems, Exercises and Suggested Reading

1. Implement the process of replication on a simple example, such as estimating the mean of a random sample from a normal distribution. Summarise the results for the various inferential formats and discuss how they agree with the statistical theory. 'Spice up' this exercise by generating the samples from a t or χ^2 distribution (with, say, 25 degrees of freedom), but pretend throughout that the samples are from a normal distribution.

2. Try to evaluate MAE of the sample mean for a random sample from the normal distribution with known variance $\sigma^2 = 1$. Describe the difficulties encountered. (There is a bonus for the complete solution.)

3. Choose a part of the simulation study in Problem 1. and repeat it to compare alternative estimators in the setting of the analysis of variance (ANOVA), described in Longford (2008).

4. Suppose $\hat{\sigma}^2$ is an unbiased estimator of the variance of a normal random sample. Is $\hat{\sigma}$ also unbiased? Let c be the 97.5-percentile of the standard normal distribution. In connection with your answer about the bias of $\hat{\sigma}$, is the coverage of the confidence interval $(\hat{\mu} - c\hat{\sigma}, \hat{\mu} + c\hat{\sigma})$ exactly 95 %? Follow up your analytical answer by simulations.

5. Discuss how model selection can be interpreted as a decision, with consequences which, at least in principle, could be carefully weighed. State what these consequences are. For every model selection procedure you use, discuss whether and how it takes such consequences into account.

6. How are finiteness of the variance and of the bias (or expectation) related?

7. Discuss (and assess) the behaviour of and the risks run by a person who tries to avoid the payment for a ride on a city tram or bus. How might he alter his behaviour by information about the (potential) fine, the frequency of inspection,

and the (legal) powers of an inspector? What other factors might influence his behaviour?

8. Discuss the motives for buying a ticket in a commercial lottery, when we know that the total payout is always smaller than the receipts. Construct some plausible loss functions for the two actions (to play and not to play) that an objector to the lottery, an enthusiastic player and somebody ambivalent (e.g., an occasional player) might have.

9. On a lighter note. Discuss the two options a pedestrian has when crossing a busy road, together with the consequences of the two kinds of bad decision that he or she could make. Discuss in the class what loss functions would reflect your preferences and how they would affect your conduct (choice of an option). How would you incorporate in the decision the information about how busy the road is?

10. A short story about a decision without studying the consequences: Longford (2007). See also Longford (2005).

11. Suggested reading about decision theory: Savage (1951); Raiffa and Schlaifer (1961); Ferguson (1969); Berger (1985); Lindley (1985) and DeGroot (2004).

References

Berger, J. O. (1985). *Statistical decision theory and bayesian analysis* (2nd ed.). New York: Springer-Verlag.

DeGroot, M. H. (2004). *Optimal statistical decisions*. New York: McGraw-Hill.

Ferguson, T. S. (1969). *Mathematical statistics: A decision theoretical approach*. New York: Academic Press.

Lindley, D. V. (1985). *Making decisions* (2nd ed.). Chichester, UK: Wiley.

Neyman, J., & Pearson, E. (1933). On the problem of the most efficient tests of statistical hypotheses. *Philosophical Transactions of the Royal Society of London Series A, 231,* 289–337.

Longford, N. (2005). Editorial: Model selection and efficiency. Is 'Which model..?' the right question? *Journal of the Royal Statistical Society Series A, 168,* 469–472.

Longford, N. T. (2007). Playing consequences [Letter to the Editor]. *Significance, 4,* 46.

Longford, N. T. (2008). An alternative analysis of variance. *SORT, Journal of the Catalan Institute of Statistics, 32,* 77–91.

R Development Core Team (2009). *R: A language and environment for statistical computing.* Vienna, Austria: R Foundation for Statistical Computing.

Raiffa, H., & Schlaifer, R. (1961). *Applied statistical decision theory*. Boston, MA: Harvard Business School.

Savage, L. J. (1951). The theory of statistical decision. *Journal of the American Statistical Association, 46,* 55–67.

Chapter 2
Estimating the Mean

This chapter deals with one of the elementary statistical problems, estimating the mean of a random sample from a normal distribution. We assume that the variance of this distribution is known. More general versions of this problem are addressed in later chapters.

Let $X_1, \ldots, X_n$ be a random sample from a normal distribution with unknown expectation μ and variance known to be equal to unity. We write $X_i \sim \mathcal{N}(\mu, 1)$, $i = 1, \ldots, n$, independently. Without a careful description of the task related to μ, we would not contemplate any estimator other than the sample mean $\hat{\mu} = (X_1 + \cdots + X_n)/n$. It is unbiased and efficient for μ; its sampling variance is equal to $1/n$. However, if we are averse to positive estimation errors, $\hat{\mu} > \mu$, then an estimator $\hat{\mu} - c$, where c is a positive constant, may be more suitable.

2.1 Estimation with an Asymmetric Loss

Suppose we associate the estimation error $\tilde{\mu} - \mu$ of an estimator $\tilde{\mu}$ of the target μ with loss $(\tilde{\mu} - \mu)^2$ when $\tilde{\mu} < \mu$, but for positive estimation error, when $\tilde{\mu} > \mu$, the loss is $R(\tilde{\mu} - \mu)^2$, where R is a constant greater than unity. This loss, as a function of $\tilde{\mu}$ and μ, is a piecewise quadratic loss function. In fact, the function depends on $\tilde{\mu}$ and μ only through the estimation error $\tilde{\mu} - \mu$. Figure 2.1 displays examples of piecewise quadratic loss functions for two values of the penalty ratio R, each with three values of the target μ. In the left-hand panel, six functions $L(\tilde{\mu}, \mu)$ are drawn, but they correspond to only two distinct functions of the error $\tilde{\mu} - \mu$ in the right-hand panel.

We explore estimators $\tilde{\mu} = \hat{\mu} - c$, where $\hat{\mu}$ is the sample mean and c is a constant that we would set. So, $\tilde{\mu} \sim \mathcal{N}(\mu - c, 1/n)$. The expected loss of $\tilde{\mu}$ is

N. T. Longford, *Statistical Decision Theory*,
SpringerBriefs in Statistics, DOI: 10.1007/978-3-642-40433-7_2,
© The Author(s) 2013

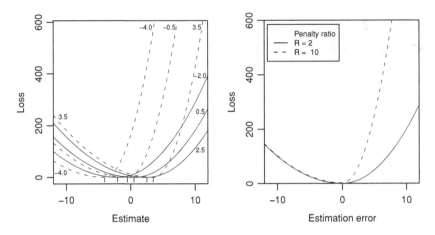

Fig. 2.1 Piecewise quadratic loss functions, as functions of estimate and target *(left-hand panel)* and of the estimation error *(right-hand panel)*. The values of the target μ are indicated in the *left-hand panel*

$$Q = R\sqrt{n} \int_{\mu}^{+\infty} (y - \mu)^2 \phi\{\sqrt{n}(y - \mu + c)\} \, dy$$

$$+ \sqrt{n} \int_{-\infty}^{\mu} (y - \mu)^2 \phi\{\sqrt{n}(y - \mu + c)\} \, dy, \qquad (2.1)$$

where ϕ is the density of the standard normal distribution, $\mathcal{N}(0, 1)$,

$$\phi(y) = \frac{1}{\sqrt{2\pi}} \exp\left(-\tfrac{1}{2}y^2\right).$$

Denote by Φ the distribution function of $\mathcal{N}(0, 1)$. The transformation $z = \sqrt{n}(y - \mu + c)$ yields the equivalent expression

$$Q = R \int_{c\sqrt{n}}^{+\infty} \left(\frac{z}{\sqrt{n}} - c\right)^2 \phi(z) \, dz + \int_{-\infty}^{c\sqrt{n}} \left(\frac{z}{\sqrt{n}} - c\right)^2 \phi(z) \, dz.$$

Denote the two terms by Q_+ and Q_-. For $R = 1$, we would obtain

$$Q_+ + Q_- = \text{MSE}(\tilde{\mu}; \mu) = c^2 + \frac{1}{n},$$

and $c = 0$ would be the optimal choice. For $R \neq 1$, a similar reduction does not take place. We work out the details for Q_-; Q_+ is dealt with similarly. By expanding the square in the integrand, we obtain

$$Q_- = \frac{1}{n} \int_{-\infty}^{c\sqrt{n}} z^2 \phi(z)\,dz - \frac{2c}{\sqrt{n}} \int_{-\infty}^{c\sqrt{n}} z\phi(z)\,dz + c^2 \int_{-\infty}^{c\sqrt{n}} \phi(z)\,dz .$$

It is easy to check that $\phi'(z) = -z\phi(z)$, so $-\phi(z)$ is a primitive function for $z\phi(z)$. The first integral is evaluated by parts, differentiating z and integrating $z\phi(z)$:

$$Q_- = \frac{1}{n} \left\{ \left[-z\phi(z) \right]_{-\infty}^{c\sqrt{n}} + \int_{-\infty}^{c\sqrt{n}} \phi(z)dz \right\} - \frac{2c}{\sqrt{n}} \left[-\phi(z) \right]_{-\infty}^{c\sqrt{n}} + c^2 \Phi\left(c\sqrt{n}\right)$$

$$= -\frac{c}{\sqrt{n}} \phi\left(c\sqrt{n}\right) + \frac{1}{n} \Phi\left(c\sqrt{n}\right) + \frac{2c}{\sqrt{n}} \phi\left(c\sqrt{n}\right) + c^2 \Phi\left(c\sqrt{n}\right)$$

$$= \left(c^2 + \frac{1}{n} \right) \Phi\left(c\sqrt{n}\right) + \frac{c}{\sqrt{n}} \phi\left(c\sqrt{n}\right) .$$

By similar steps we obtain the identity

$$Q_+ = R\left(c^2 + \frac{1}{n} \right) \left\{ 1 - \Phi\left(c\sqrt{n}\right) \right\} - \frac{cR}{\sqrt{n}} \phi\left(c\sqrt{n}\right) . \tag{2.2}$$

Hence the expected loss $Q = Q_+ + Q_-$ is

$$Q = \left(c^2 + \frac{1}{n} \right) \left\{ R - (R-1) \Phi\left(c\sqrt{n}\right) \right\} - \frac{c(R-1)}{\sqrt{n}} \phi\left(c\sqrt{n}\right) . \tag{2.3}$$

2.2 Numerical Optimisation

It remains to find the constant c for which the expected loss Q in (2.3) is minimised. This cannot be done by a closed-form expression. We apply the Newton-Raphson algorithm. It is an iterative procedure which generates a provisional (approximate) solution $c^{(i+1)}$ in iteration $i+1$ by adjusting the previous solution $c^{(i)}$. The adjustment depends on the first- and second-order derivatives of Q with respect to c:

$$c^{(i+1)} = c^{(i)} - \frac{s\left(c^{(i)}\right)}{H\left(c^{(i)}\right)}, \tag{2.4}$$

where $s = \partial Q/\partial c$ and $H = \partial^2 Q/\partial c^2$ are treated as functions of c. They are called the score and the Hessian (functions), respectively. The iterations are stopped when the absolute value of the adjustment $c^{(i+1)} - c^{(i)} = -s/H$, or of the score s, becomes smaller than a prescribed small quantity, such as 10^{-8}.

We derive the adjustment (2.4) to gain an understanding of the properties of this algorithm and to formulate its assumptions. Obviously, the first- and second-order derivatives of Q have to exist in the range of plausible values of c. This is satisfied for

Q in (2.3). The Taylor expansion for the first-order derivative at the exact solution c^*, centred around the current (provisional) solution $c^{(i)}$, is

$$s\left(c^*\right) \doteq s\left(c^{(i)}\right) + \left(c^* - c^{(i)}\right) H\left(c^{(i)}\right). \tag{2.5}$$

At the minimum of Q, $s(c^*) = 0$. Regarding (2.5) as an identity, setting aside the fact that it is merely an approximation, and solving it for c^*, we obtain the updating formula in (2.4). Hopefully this gets us closer to c^*. From the derivation, it is clear that this algorithm converges fast when the approximation in (2.5) is precise, that is, when the function s is close to linearity—when Q is close to a quadratic function, or when a solution $c^{(i)}$ is already close to c^*. Problems arise when H is not a smooth function and the values of $1/H$ are not changing at a sedate pace, or indeed when $H = 0$. If iterations reach a region where $H(c) \doteq 0$, the consecutive values $c^{(i)}$ may become unstable. The Newton-Raphson iterations require an initial solution $c^{(0)}$. It can be set by trial and error if we have only one problem to solve. For finding the minimum of Q, $c^{(0)} = 0$ is a suitable initial solution.

When it converges, the Newton-Raphson algorithm finds a root of the score function. The function s may have several roots and the one we find may not be a (global) minimum of Q. However, if s is an increasing function, then the root is unique and it is the only minimum of Q. Often a simple way of proving that s is increasing is by checking that H is positive at the root (or throughout).

For the function in (2.3), we have

$$
\begin{aligned}
s \;=\; \frac{\partial Q}{\partial c} &= 2c\left\{R - (R-1)\,\Phi\left(c\sqrt{n}\right)\right\} - \left(c^2 + \frac{1}{n}\right)\sqrt{n}(R-1)\phi\left(c\sqrt{n}\right) \\
&\quad - \frac{R-1}{\sqrt{n}}\phi\left(c\sqrt{n}\right) + c^2\sqrt{n}(R-1)\phi\left(c\sqrt{n}\right) \\
&= 2cR - 2(R-1)\left\{c\,\Phi\left(c\sqrt{n}\right) + \frac{\phi\left(c\sqrt{n}\right)}{\sqrt{n}}\right\}
\end{aligned}
$$

$$
\begin{aligned}
H \;=\; \frac{\partial^2 Q}{\partial c^2} &= 2R - 2(R-1)\left\{\Phi\left(c\sqrt{n}\right) - c\sqrt{n}\,\phi\left(c\sqrt{n}\right) + c\sqrt{n}\,\phi\left(c\sqrt{n}\right)\right\} \\
&= 2\left\{R - (R-1)\,\Phi\left(c\sqrt{n}\right)\right\}. \tag{2.6}
\end{aligned}
$$

From this we conclude that $2 < H(c) < 2R$ for all $R > 0$ (not only for $R > 1$), so Q has a unique minimum, and it is at the root of s. Since $H(c) > 2$, there are no convergence problems.

By way of an example, suppose $n = 50$ and $R = 20$. We set the initial solution to $c^{(0)} = 0$; the corresponding value of Q is $(R + 1)/(2n) = 0.21$. The progression of the provisional solutions is displayed in Table 2.1. The right-most column (Precision) is defined as

$$-\frac{1}{2}\log_{10}\left[\left(c^{(i)} - c^{(i-1)}\right)^2 + \left\{s\left(c^{(i)}\right)\right\}^2\right] \tag{2.7}$$

Table 2.1 Iterations of the Newton-Raphson algorithm to minimise the expected loss Q with penalty ratio $R = 20$ and size $n = 50$ of a random sample from $\mathcal{N}(\mu, 1)$; piecewise quadratic loss

Iteration (i)	$c^{(i)}$	$Q^{(i)}$	Precision
0	0.0000	0.2100	
1	0.1021	0.2100	−0.33
2	0.1511	0.0820	0.27
3	0.1632	0.0674	1.04
4	0.1639	0.0668	2.36
5	0.1639	0.0668	4.93
6	0.1639	0.0668	10.08

(logarithm with base 10). It can be interpreted as the number of digits of precision. The iterations are stopped when this quantity exceeds 8.0. The table indicates that convergence is achieved after six iterations, although we could have stopped after just four. However, the calculations, done in R, are instant, so the additional two iterations represent no waste of our resources.

Thus, the estimator with the minimum expected loss when $n = 50$ and $R = 20$ is $\hat{\mu} - 0.1639$ and the corresponding expected loss is 0.0668. The expected loss with the unbiased estimator is 0.2100, more than three times greater. It is easy to show that when $\sigma^2 \neq 1$, $\hat{\mu} - 0.1639\sigma$ is the estimator with the smallest expected loss.

2.3 Plausible Loss Functions

In practice, it is difficult to set the penalty ratio R to a single value without leaving some doubt that R may be somewhat greater or smaller. We regard this as a source of uncertainty associated with the *elicitation* process, the dialogue between the analyst and the client, in which the background and details of the problem are discussed. We address it by solving the problem for a range of values of R that were agreed to be plausible. A range of penalty ratios (R_L, R_U), and any value within it, is said to be plausible if the client would rule out any value of R outside this range. At the same time, the plausible range should be set to as narrow an interval as possible.

Figure 2.2 presents the continuum of solutions c^* and the corresponding expected losses for sample sizes $10 \leq n \leq 200$ and penalty ratios $5 \leq R \leq 100$. Denote these functions (curves) by $c_R(n)$ and $Q_R(n)$, respectively. The panels at the top plot c^* and Q as functions of n on the linear (original) scale, and the panels at the bottom reproduce them on the log scale for n. The log scale is useful because at the planning stage one is more likely to consider increasing or reducing the sample size by a certain multiple, such as 1.25 or (25 %), and that corresponds to an increase or reduction by a constant, $\log(1.25)$, on the log scale.

The diagram shows that the optimal shift c^* is positive throughout, it increases with R and decreases with n, steeply for small n. For small sample sizes, the functions $c_R(n)$ and $Q_R(n)$ have steep gradients on R, so a lot is at stake. For large sample

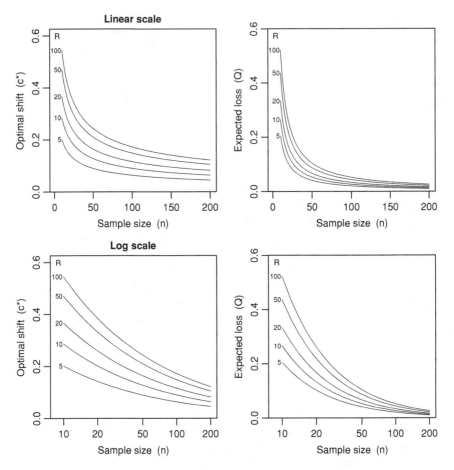

Fig. 2.2 The offset c^* for which the estimator $\hat{\mu} - c^*$ has minimum expected loss, as a function of the sample size n and the penalty ratio R; piecewise quadratic loss. The corresponding expected loss is plotted in the *right-hand panels*

sizes, the differences diminish. In fact, for any fixed R, $c_R(n)$ converges to zero as $n \to +\infty$, but the convergence is rather slow.

Some of these conclusions can be confirmed directly from (2.6). Since H is positive, s is an increasing function of c. But $s(0) = -2(R-1)\phi(0)/\sqrt{n} < 0$, so c^*, the root of s, has to be positive. Further, $s\sqrt{n}$ depends on c and $\sqrt{n}$ only through $c\sqrt{n}$. If c_1 is the root of s for n_1, then $c_2 = c_1\sqrt{n_1/n_2}$ is the root for n_2. Therefore the root of s, which coincides with the root of $s\sqrt{n}$, is a decreasing function of n and $c_R(n) \to 0$ as $n \to +\infty$. Similarly $s\sqrt{n}/(R-1)$ is a decreasing function of R, so long as $c > 0$ and $R > 1$:

$$\frac{s(c)\sqrt{n}}{R-1} = \frac{2c\sqrt{n}R}{R-1} - 2\left\{c\sqrt{n}\,\Phi\left(c\sqrt{n}\right) + \phi\left(c\sqrt{n}\right)\right\}.$$

By definition, $s\{c_R(n)\} = 0$. By substituting $R' > R$ for R and $c_R(n)$ for c on the right-hand side, we obtain a negative quantity. Since s is increasing, $c_{R'}(n)$ has to be greater than $c_R(n)$.

Each curve in Fig. 2.2 is drawn by connecting the values of $c_R(n)$ and $Q_R(n)$ for a fine grid of values n. We set $n = 10, 12, \ldots, 200$. A coarser grid may suffice, but the saving in the computing involved is insubstantial. In fact, all the evaluations for the diagram took only 0.19 sec. of CPU time on a Mac laptop. In R, a function is declared, with arguments n, R, and some others that specify the convergence criterion and control the output. One output has the format of Table 2.1, with the details of the iterations, and the other gives only the 'bottom line': c^*, $Q(c^*)$ and the number of iterations. For the evaluations presented in Fig. 2.2, between five and eight iterations were required. The function returns the results for one setting of n and R, but its repeated application for a range of values of n and R requires minimum programming effort, using the system-defined function `apply`.

When the second-order derivative is not available, or we want to avoid its evaluation because it is too complex, the Newton (linearisation) method can be applied. In this method, a pair of provisional solutions, (c_A, c_B), defines the following approximation to the root of s:

$$c_D = c_A - \frac{c_B - c_A}{s(c_B) - s(c_A)} s(c_A).$$

This rule is applied iteratively. In the next iteration, the pair (c_B, c_D) is used in place of (c_A, c_B). The iterations are stopped when the two provisional solutions are very close to one another and the value of s for both of them is sufficiently close to zero. A criterion similar to (2.7) can be formulated.

2.4 Other Classes of Loss Functions

In this section, we extend the repertoire of loss functions for which estimation with minimum expected loss is tractable.

The piecewise linear loss for estimator $\hat\theta$ of θ is defined as $\theta - \hat\theta$ when $\hat\theta < \theta$ and as $R(\hat\theta - \theta)$ when $\hat\theta > \theta$. The penalty ratio $R > 0$ plays a role similar to its namesake for piecewise quadratic loss, to reflect the aversion to positive estimation errors (when $R > 1$). The expected loss of an estimator $\tilde\mu = \hat\mu - c$ of the mean of the normal distribution with unit variance is

$$R\sqrt{n} \int_\mu^{+\infty} (y - \mu)\phi\left\{\sqrt{n}(y - \mu + c)\right\} dy$$
$$+ \sqrt{n} \int_{-\infty}^\mu (\mu - y)\phi\left\{\sqrt{n}(y - \mu + c)\right\} dy;$$

compare with (2.1). By steps similar to those used in deriving (2.3), we obtain

$$Q = \frac{R}{\sqrt{n}}\left[-\phi(z)\right]_{c\sqrt{n}}^{+\infty} - cR\left\{1 - \Phi(c\sqrt{n})\right\} + c\,\Phi(c\sqrt{n}) - \frac{1}{\sqrt{n}}\left[-\phi(z)\right]_{-\infty}^{c\sqrt{n}}$$

$$= (R+1)\left\{c\,\Phi(c\sqrt{n}) + \frac{\phi(c\sqrt{n})}{\sqrt{n}}\right\} - cR.$$

Its derivatives with respect to c are

$$s = (R+1)\,\Phi(c\sqrt{n}) - R$$
$$H = (R+1)\sqrt{n}\,\phi(c\sqrt{n}),$$

simpler than for the piecewise quadratic loss. We have a closed-form solution for minimising Q,

$$c^* = \frac{1}{\sqrt{n}}\,\Phi^{-1}\left(\frac{R}{R+1}\right). \tag{2.8}$$

The discussion of the properties of this solution is left for an exercise.

In principle, any loss function can be declared that is increasing in the absolute estimation error $\Delta = |\tilde{\mu} - \mu|$ and for which $L(0) = 0$. The latter condition is not important, because we could adjust L as $L - L(0)$; we only need $L(0)$ to be well defined. A reasonable condition is that L be continuous, although it does not have to be differentiable throughout. Apart from additivity (Sect. 1.7), the key criterion of usefulness of a loss function is that it reflects the client's perspective. The following example shows, however, that some loss functions lead to unreasonable answers.

The piecewise constant loss function is defined as the constant unity for negative estimation error and $R > 0$ for positive estimation error. The expected loss for estimating μ by $\tilde{\mu} = \hat{\mu} - c$ is

$$Q = R\sqrt{n}\int_{\mu}^{+\infty} \phi\{\sqrt{n}(y - \mu + c)\}\,dy + \sqrt{n}\int_{-\infty}^{\mu} \phi\{\sqrt{n}(y - \mu + c)\}\,dy$$
$$= R - (R-1)\,\Phi(c\sqrt{n}),$$

which has no minimum, but suggests the solution $c^* = +\infty$, that is, the 'estimator' $\tilde{\mu} = -\infty$. Since $P(\tilde{\mu} \neq \mu) = 1$, we are certain to pay a penalty. We should therefore avoid positive estimation error (with penalty $R > 1$), and that is achieved with a sufficiently small estimate (large c). Thus, reducing our attention to the sign of the estimation error is a bad strategy; its size also matters.

2.4.1 LINEX Loss

The LINEX loss for estimation error $\Delta = \hat{\theta} - \theta$ is defined as

$$L_a(\Delta) = \exp(a\Delta) - a\Delta - 1;$$

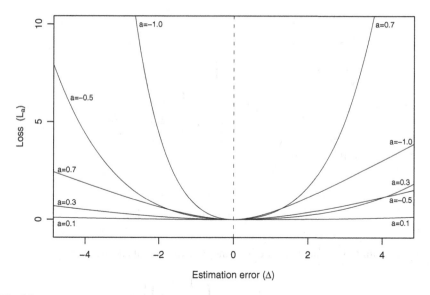

Fig. 2.3 Examples of LINEX loss functions

$a \neq 0$ is a constant to be set. It is easy to check that L_a has all the attributes of a loss function: $L_a(0) = 0$ and $L_a(\Delta)$ decreases for negative Δ and increases for positive Δ. The function is drawn in Fig. 2.3 for a few coefficients a.

For $x \gg 0$ (x positive and large), $\exp(x) \gg x+1$, so when a and Δ have the same sign and $a\Delta$ is large, $(a\Delta+1)/L_a(\Delta) \doteq 0$ and $L_a(\Delta)$ behaves similarly to $\exp(a\Delta)$. In contrast, when a and Δ have opposite signs and $-a\Delta$ is large, $\exp(a\Delta)/L_a(\Delta) \doteq 0$, and $L_a(\Delta) \ll 1 - a\Delta \ll L_a(-\Delta)$. So, L_a is distinctly asymmetric, with greater values for large positive errors when $a > 0$, and greater values for large negative errors when $a < 0$.

The expected LINEX loss of $\hat{\mu} - c$ is

$$Q_a = \sqrt{n} \int_{-\infty}^{+\infty} L_a(y - \mu) \, \phi\{\sqrt{n}\,(y - \mu + c)\} \, \mathrm{d}y$$

$$= \int_{-\infty}^{+\infty} \exp\left(\frac{az}{\sqrt{n}} - ac\right) \phi(z) \, \mathrm{d}z - \int_{-\infty}^{+\infty} \left(\frac{az}{\sqrt{n}} - ac\right) \phi(z) \, \mathrm{d}z - 1.$$

The latter integral is equal to $-ac$ because ϕ is symmetric and it integrates to unity. The former integral can be related to the expectation of a lognormal distribution. If $X \sim \mathcal{N}(\mu, \sigma^2)$, then $\mathrm{E}\{\exp(X)\} = \exp(\mu + \frac{1}{2}\sigma^2)$. To make the text self-contained, we derive it from basic principles.

By consolidating the arguments of the exponentials and matching the result with a normal density, $\mathcal{N}(a/\sqrt{n}, 1)$, we obtain

$$\int_{-\infty}^{+\infty} \exp\left(\frac{az}{\sqrt{n}} - ac\right) \phi(z)\, dz$$

$$= \frac{1}{\sqrt{2\pi}} \exp(-ac) \int_{-\infty}^{+\infty} \exp\left(-\frac{z^2}{2} + \frac{az}{\sqrt{n}}\right) dz$$

$$= \frac{1}{\sqrt{2\pi}} \exp\left(\frac{a^2}{2n} - ac\right) \int_{-\infty}^{+\infty} \exp\left\{-\frac{1}{2}\left(z - \frac{a}{\sqrt{n}}\right)^2\right\} dz$$

$$= \exp\left(\frac{a^2}{2n} - ac\right).$$

Hence

$$Q_a = \exp\left(\frac{a^2}{2n} - ac\right) + ac - 1.$$

The minimum of this function of c is found by exploring its derivative:

$$s_a = a\left\{1 - \exp\left(\frac{a^2}{2n} - ac\right)\right\}.$$

Further differentiation yields the Hessian

$$H_a = a^2 \exp\left(\frac{a^2}{2n} - ac\right).$$

Since $H_a > 0$, Q_a has a unique minimum, and it is at the root of s_a. The root is $c^* = a/(2n)$ and the minimum attained is $Q_a(c^*) = ac^* = a^2/(2n)$. The expected loss with $\hat{\mu}$, which corresponds to $c = 0$, is $\exp\{a^2/(2n)\} - 1$. The difference of the losses, $\exp\{a^2/(2n)\} - a^2/(2n) - 1$, is equal to the loss $L_1\{a^2/(2n)\} = L_a\{a/(2n)\}$. The expected loss decreases with n to zero, but for small to moderate n it is substantial, especially when $|a|$ is large.

2.5 Comparing Two Means

In this section we address the problem of deciding which of two random samples from normal distributions with the identical variances is greater. We assume that the common variance, σ^2, is known. No generality is lost by assuming that $\sigma^2 = 1$, because we can reformulate the problem for samples x_1 and x_2 as a problem for $\sigma^{-1}x_1$ and $\sigma^{-1}x_2$. Denote the expectations of the two samples by μ_1 and μ_2 and set $\Delta = \mu_2 - \mu_1$. Let n_1 and n_2 be the sizes of the two samples and $\hat{\Delta} = \hat{\mu}_2 - \hat{\mu}_1$ the difference of the sample means. Its distribution is $\mathcal{N}(\Delta, m\sigma^2)$, where $m = 1/n_1 + 1/n_2$; $1/m$ can be interpreted as the effective sample size of the pair of the samples.

With hypothesis testing, we set the size of the test, α, by convention to 0.05, although other choices (probabilities) are permitted, and choose the critical region, denoted by $\mathcal{C}$, such that under the (null) hypothesis that $\Delta = 0$ the probability that a new realisation of $\widehat{\Delta}$ falls within $\mathcal{C}$ is equal to α. Common choices for $\mathcal{C}$ are the complement of a symmetric interval, $\{-\infty, \sigma\sqrt{m}\,\Phi^{-1}(\frac{1}{2}\alpha)\} \cup \{\sigma\sqrt{m}\,\Phi^{-1}(1 - \frac{1}{2}\alpha), +\infty\}$, and the one-sided intervals $\{\sigma\sqrt{m}\,\Phi^{-1}(1 - \alpha), +\infty\}$ and $\{-\infty, \sigma\sqrt{m}\,\Phi^{-1}(\alpha)\}$. If $\widehat{\Delta} \in \mathcal{C}$, we reject the null hypothesis. Otherwise, we have no evidence against the null hypothesis. Interpreting the latter outcome as a confirmation that $\Delta = 0$, or that $|\Delta|$ is small, is not appropriate. Following it up by action that would be apropriate if $\Delta = 0$ but not otherwise, has no logical basis.

Suppose we have a research or business agenda the details of which depend on Δ. If we knew that $\Delta < 0$, action A would be appropriate. Otherwise we would pursue action B. If we elect action A but $\Delta > 0$, we incur loss μ^2; if we elect action B but $\Delta < 0$, we lose $R\mu^2$. Note that this loss function differs from the function of the same name defined in Sect. 2.1, because no loss is incurred when the correct sign is chosen, even if $\widehat{\Delta}$, or another estimate, differs substantially from Δ. Because of the symmetry of the problem, we lose no generality by assuming that $R > 1$, so that its label, *penalty* ratio, is well motivated.

We intend to base the choice between A and B on $\widehat{\Delta} - c$, where c is a constant to be set by the criterion of minimum expected loss. Since $(\widehat{\Delta} - \Delta)/\sqrt{m}$ has the standard normal distribution, $\mathcal{N}(0, 1)$, we can represent Δ by a random variable $\widehat{\Delta}+\delta$, where $\delta \sim \mathcal{N}(0, m)$. Note that we rely in this on the symmetry of $\mathcal{N}(0, m)$. Thus, we have converted an unknown constant, Δ, into a random variable, to represent our uncertainty about its value after its estimate $\widehat{\Delta}$ has been realised; that is, we converted it from a random variable to a constant. We will make these changes of status more formal in Chap. 4 within a Bayesian perspective.

When $\widehat{\Delta} - c < 0$, and so we choose action A, the expected loss is

$$Q_- = \frac{1}{\sqrt{m}} \int_0^{+\infty} x^2 \phi\left(\frac{x - \widehat{\Delta}}{\sqrt{m}}\right) dx$$
$$= \int_{-a}^{+\infty} \left(\widehat{\Delta} + z\sqrt{m}\right)^2 \phi(z)\, dz$$
$$= \widehat{\Delta}^2\{1 - \Phi(-a)\} - 2\widehat{\Delta}\sqrt{m}\left[\phi(z)\right]_{-a}^{+\infty} + m\int_{-a}^{+\infty} z^2\phi(z)\, dz\,,$$

where $a = \widehat{\Delta}/\sqrt{m}$. The latter integral is evaluated by parts,

$$\int_{-a}^{+\infty} z^2\phi(z)\, dz = \Phi(a) - a\phi(a),$$

exploiting the symmetry of the standard normal distribution, that is, $\phi(-a) = \phi(a)$ and $1 - \Phi(-a) = \Phi(a)$. Therefore,

$$Q_- = m \left\{ \left(1 + a^2\right) \Phi(a) + a\phi(a) \right\}.$$

By similar steps we obtain the expected loss when choosing action B:

$$Q_+ = mR \left[\left(1 + a^2\right) \{1 - \Phi(a)\} - a\phi(a) \right].$$

For $\widehat{\Delta}$ given, we select the action with the smaller expected loss. For small values of $\widehat{\Delta}$ action A and for large values action B is preferred. There is a critical value of $\widehat{\Delta}$ where we switch from the preference for one action to the other. This occurs at the *equilibrium*, where $Q_- = Q_+$. To prove that there is a unique equilibrium, we show that Q_- is increasing and Q_+ is decreasing. The derivatives of these functions of a are

$$\frac{\partial Q_-}{\partial a} = 2m \{a\Phi(a) + \phi(a)\}$$

$$\frac{\partial Q_+}{\partial a} = 2mR \left[a \{1 - \Phi(a)\} - \phi(a) \right].$$

Both derivatives, as functions of a, are increasing because their respective derivatives, $2m\Phi(a)$ and $2mR\{1 - \Phi(a)\}$, are positive. Whereas $\partial Q_- / \partial a$ is positive, because its limits at $\pm\infty$ are zero and $+\infty$, $\partial Q_+ / \partial a < 0$, because its limits are $-\infty$ and zero. Therefore Q_- is increasing and Q_+ is decreasing throughout $(-\infty, +\infty)$. So, our best bet is to set the constant c at the equilibrium, where $Q_- = Q_+$. This condition is

$$\Delta Q = (R + 1) \left\{ \left(1 + a^2\right) \Phi(a) + a\phi(a) \right\} - R(1 + a^2) = 0, \qquad (2.9)$$

with the factor m dropped. It is solved by the Newton-Raphson algorithm, using the expression

$$\frac{\partial \Delta Q}{\partial a} = 2(R + 1) \{a \Phi(a) + \phi(a)\} - 2aR.$$

For the solution a^*, the optimal constant c is $c^* = a^* \sqrt{m}$. The decision about the sign of Δ is based on the sign of $\widehat{\Delta} - a^* \sqrt{m}$. It is rather fortuitous that (2.9) involves the sample sizes n_1 and n_2 only through m, and m only through a. Therefore, it is practical to solve (2.9) for a range of values of R, and then convert the solution a_R^* to $c_R = a_R^* \sqrt{m}$.

For the piecewise linear loss, we find the equilibrium by evaluating the two parts of the expected loss:

$$Q_- = \frac{1}{\sqrt{m}} \int_0^{+\infty} x \phi\left(\frac{x - \widehat{\Delta}}{\sqrt{m}}\right) dx$$

$$= \widehat{\Delta}\{1 - \Phi(-a)\} - \left[\phi(z)\right]_{-a}^{+\infty}$$

$$= \sqrt{m} \{a\Phi(a) + \phi(a)\}$$

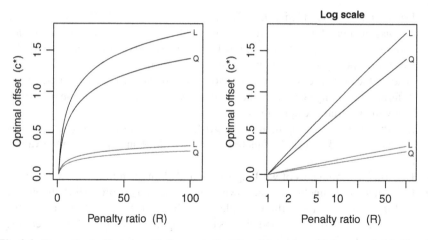

Fig. 2.4 The optimal offsets c_R with the quadratic (Q) and linear loss (L) for $m = 1$ *(black)* and $m = 1/5$ *(gray)*, as functions of the penalty ratio R, on the linear and log scales

and

$$Q_+ = R\sqrt{m}\,\{a\Phi(a) + \phi(a) - a\}.$$

Hence the balance equation

$$\Delta Q = (R - 1)\,\{a\Phi(a) + \phi(a)\} - Ra = 0,$$

which is solved by the Newton-Raphson algorithm, in which we use the identity $\partial \Delta Q / \partial a = (R - 1)\Phi(a) - R$.

Figure 2.4 displays the solutions c_R for the linear and quadratic loss functions with penalty ratios $R \in (1, 100)$ for $m = 1$ (e.g., $n_1 = n_2 = 2$, drawn in black) and $m = 0.2$ (e.g., $n_1 = n_2 = 10$, gray). The function c_R increases with R, approximately linearly on the log scale (see the right-hand panel). Of course, c^* is smaller for larger samples, in proportion of $\sqrt{m}$. Piecewise linear and quadratic loss functions are, strictly speaking, not comparable even when defined with the same penalty ratio R. However, a client may not be certain as to which of these loss functions is appropriate, so contemplating both of them is within the spirit of a wide range of plausible loss functions.

2.6 Problems, Exercises and Suggested Reading

1. Compare by simulations the sampling variances of the mean, median, the average of the two quartiles and the average of the minimum and maximum of a simple random sample from a normal distribution. Repeat this exercise with the uniform distribution on $(0, 2\theta)$ to estimate θ.

2. Derive in detail the identity in (2.2).

3. Discuss methods for finding the root of s without using its derivative H. Compare the programming effort, the results and the speed of convergence with the Newton-Raphson algorithm on examples of your choice.

4. Discuss the properties of c^* in (2.8). How is c^* adjusted when the variance σ^2 is different from unity? Compare the minimum value of Q with its value for $c = 0$. Study their difference and ratio as $n \to +\infty$.

5. Plot the values of the optimal shift $c_R(n)$ as functions of R for a selection of sample sizes n. Explore the function `contour` in R and apply it to the values of $c_R(n)$.

6. Discuss the advantages of working with n and R on the multiplicative scale.

7. Construct a loss function of your own choice based on the properties you would find desirable for a specific example or application. Search the literature for examples of loss functions and discuss their properties. Plot these loss functions, e.g., using the layout of Fig. 2.1.

8. Discuss how the loss function should be adapted for estimating a transformed parameter. For example, we may have a particular loss function for estimating the mean μ of a normal random sample (with known variance), but we wish to know the value of $\exp(\mu)$. Suggested reading about the lognormal distribution: Aitchison and Brown (1957) and Crow and Shimizu (1988). See Longford (2009) for estimating the mean and median of the lognormal distribution in small samples.

9. Show that loss functions form classes of equivalence. Two loss functions, L_1 and L_2, fall into the same class if $L_1 = bL_2$ for some scalar $b > 0$. When is a linear combination of two loss functions, $aL_1 + bL_2$, also a loss function? Construct such a loss function.

10. Discuss how the results of Sect. 2.5 can be applied to deciding whether the expectation of a normally distributed sample (with a known variance) is positive or negative.

    Hint: Suppose in the comparison of two samples, one is so large that its expectation is, in effect, known.

11. Explore estimation of μ in the context of Sect. 2.5 with piecewise constant loss.

12. The switch between the statuses of fixed and random for the parameter of interest in Sect. 2.5 is associated with the fiducial argument. See Seidenfeld (1992) for background.

13. Suggested reading about methods for optimisation: Lange (1999), Chaps. 5, 11, 13, and Press et al. (2007), Chap. 9.

14. Suggested reading about LINEX loss: Zellner (1986).

15. Suggested reading of historical interest: Friedman and Savage (1948); Wald (1950); Le Cam (1955); Pratt et al. (1964) . Also, Pratt et al. (1995).

16. An unsolved problem. Why is c^* approximately linearly related to $\log(R)$?

17. Derive the integral of Φ and the integral of the result, and explain the appearance of these functions in Sect. 2.4.

References

Aitchison, J., & Brown, J. A. C. (1957). *The lognormal distribution*. Cambridge: Cambridge University Press.

Crow, E. L., & Shimizu, K. (Eds.). (1988). *Lognormal distributions*. New York: Theory and Applications. M. Dekker.

Friedman, M., & Savage, L. J. (1948). The utility analysis of choices involving risk. *Journal of Political Economy*, *56*, 279–304.

Lange, K. (1999). *Numerical analysis for statisticians*. New York: Springer-Verlag.

Le Cam, L. (1955). An extension of Wald's theory of statistical decision functions. *Annals of Mathematical Statistics*, *26*, 69–81.

Longford, N. T. (2009). Inference with the lognormal distribution. *Journal of Statistical Planning and Inference*, *139*, 2329–2340.

Pratt, J. W., Raiffa, H., & Schlaifer, R. (1964). The foundations of decision under uncertainty: An elementary exposition. *Journal of the American Statistical Association*, *59*, 353–375.

Pratt, J. W., Raiffa, H., & Schlaifer, R. (1995). *Introduction to statistical decision theory*. Cambridge: MIT Press.

Press, W.H., Teukolsky, S.A., Vetterling, W.T., & Flannery, B.P. (2007). *Numerical recipes: The art of scientific computing (3rd ed.)*. Cambridge University Press, New York.

Seidenfeld, T. (1992). R. A. Fisher's fiducial argument and Bayes' theorem. *Statistical Science*, *7*, 358–368.

Wald, A. (1950). *Statistical decision functions*. New York: Wiley.

Zellner, A. (1986). Bayesian estimation and prediction using asymmetric loss functions. *Journal of the American Statistical Association*, *81*, 446–451.

Chapter 3
Estimating the Variance

This chapter deals with estimation of the variance of a normal distribution. We review briefly how it is dealt with by the standard curriculum, identify its weaknesses, and address them by tailoring the estimator more closely to the purpose for which it is intended.

3.1 Unbiased and Efficient Estimation

Suppose we have a random sample $X_1, \ldots, X_n$ from a normal distribution $\mathcal{N}(\mu, \sigma^2)$. We are interested in estimating the variance σ^2. The commonly used estimator is

$$\hat{\sigma}^2 = \frac{1}{n-1} \sum_{i=1}^{n} (X_i - \hat{\mu})^2, \tag{3.1}$$

where $\hat{\mu} = (X_1 + \cdots + X_n)/n$ is the sample mean. The sampling distribution of $\hat{\sigma}^2$ is related to the χ^2 distribution with $n-1$ degrees of freedom as

$$(n-1) \frac{\hat{\sigma}^2}{\sigma^2} \sim \chi_{n-1}^2.$$

We say that $\hat{\sigma}^2$ has a scaled χ^2 distribution with $n-1$ degrees of freedom. The density of the χ^2 distribution with k degrees of freedom is

$$f_k(x) = \frac{1}{2\,\Gamma_2(k)} \left(\tfrac{1}{2}x\right)^{k/2-1} \exp\left(-\tfrac{1}{2}x\right), \tag{3.2}$$

where Γ_2 denotes the half-gamma function, $\Gamma_2(k) = \Gamma(\tfrac{1}{2}k)$. We introduce it solely for typographical reasons. The χ_k^2 distribution has expectation k and variance $2k$. Scaling refers to a scalar multiple, in our case, $\sigma^2/(n-1)$.

N. T. Longford, *Statistical Decision Theory*,
SpringerBriefs in Statistics, DOI: 10.1007/978-3-642-40433-7_3,
© The Author(s) 2013

Using n instead of the divisor $n - 1$ in (3.1) is generally regarded as a mistake because $\hat{\sigma}^2$ is unbiased for σ^2. However, the efficient estimator of σ^2 is $\hat{\sigma}^2(n - 1)/(n + 1)$ — we should use the divisor $n + 1$ instead of $n - 1$. We prove this by searching for the constant c for which $c\hat{\sigma}^2$ has minimum MSE. We have

$$\text{MSE}\left(c\hat{\sigma}^2; \sigma^2\right) = \sigma^4 \left\{ \frac{2c^2}{n - 1} + (c - 1)^2 \right\}.$$

This quadratic function of c attains its unique minimum at $c^* = (n - 1)/(n + 1)$, equal to $2\sigma^4/(n + 1)$. The MSE of the unbiased estimator is $2\sigma^4/(n - 1)$.

Variances and their estimators tend to be easy to handle analytically, but for interpretation we sometimes prefer the standard deviation σ, because it is defined on the same scale as the original observations X_i. However, $\hat{\sigma}$, the obvious choice for estimating σ, is neither unbiased nor efficient. The expectation of $\hat{\sigma}$ is

$$
\begin{aligned}
\text{E}\left(\hat{\sigma}\right) &= \frac{\sigma}{\sqrt{n - 1}} \int_0^{+\infty} \frac{\sqrt{x}}{2\,\Gamma_2(n - 1)} \left(\frac{x}{2}\right)^{(n-1)/2-1} \exp\left(-\frac{x}{2}\right) dx \\
&= \frac{\sigma}{\sqrt{2(n - 1)}} \frac{1}{\Gamma_2(n - 1)} \int_0^{+\infty} \left(\frac{x}{2}\right)^{n/2-1} \exp\left(-\frac{x}{2}\right) dx \\
&= \frac{\sigma\sqrt{2}}{\sqrt{n - 1}} \frac{\Gamma_2(n)}{\Gamma_2(n - 1)}.
\end{aligned}
\tag{3.3}
$$

The last identity is obtained by relating the integrand to the χ_n^2 distribution. We avoid references to a complex expression by writing $\text{E}(\hat{\sigma}) = H_n\sigma$; H_n depends only on n. It is well defined also for non-integer values of n greater than 1.0, so we can treat it as a function of n. It is plotted in the left-hand panel of Fig. 3.1. The function

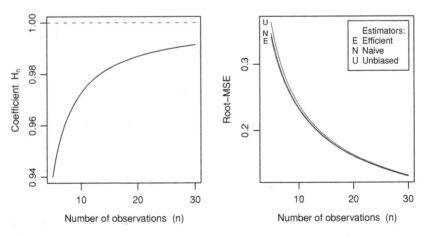

Fig. 3.1 The coefficient H_n as a function of n and the root-MSEs of the alternative estimators of σ

increases throughout $n \in (1, +\infty)$ and converges to unity as $n \to +\infty$. The bias of $\hat{\sigma}$ is $\sigma(H_n - 1)$. The variance of $\hat{\sigma}$ is

$$\mathrm{var}(\hat{\sigma}) = \mathrm{E}\left(\hat{\sigma}^2\right) - \left\{\mathrm{E}\left(\hat{\sigma}\right)\right\}^2$$
$$= \sigma^2 \left(1 - H_n^2\right).$$

For a positive constant c, the estimator $c\hat{\sigma}$ has MSE

$$\mathrm{MSE}\left(c\hat{\sigma}; \sigma\right) = \sigma^2 \left\{c^2\left(1 - H_n^2\right) + (1 - cH_n)^2\right\}$$
$$= \sigma^2 \left(1 - 2cH_n + c^2\right).$$

Therefore the efficient estimator (in the class $c\hat{\sigma}$) is $H_n\hat{\sigma}$. The root-MSE of this estimator is $\sigma\sqrt{1 - H_n^2}$. The right-hand panel of Fig. 3.1 compares it to the root-MSE of the naive estimator $\hat{\sigma}$, which is $\sigma\sqrt{2(1 - H_n)}$, and the root-MSE of the unbiased estimator $\hat{\sigma}/H_n$, which is $\sigma\sqrt{1/H_n^2 - 1}$, for $\sigma^2 = 1$ and $5 \leq n \leq 30$. The root-MSE scale is better suited for a display, because the values are greater and are on the same scale as the original observations. Arguably, the differences among the alternative estimators are small, except perhaps for the smallest sample sizes, but the improvement is obtained by very little additional computing. The unbiased estimator is the least efficient of the three.

3.2 Loss Functions for Variance Estimators

Adopting the view that overestimation of σ^2 is not as serious as underestimation by the same quantity, we consider estimating σ^2 with asymmetric loss functions. We choose piecewise linear loss and explore the estimators $\tilde{\sigma}^2 = c\hat{\sigma}^2$ with positive constants c. We prefer the linear loss to the quadratic loss, because σ^2 and $\tilde{\sigma}^2$ already entail a squaring when related to σ and $\tilde{\sigma}$, or to the observations X, for which we might consider piecewise quadratic loss. As an aside, we mention that quadratic loss for $\tilde{\sigma}$ does not coincide with linear loss for $\tilde{\sigma}^2$, since

$$\tilde{\sigma}^2 - \sigma^2 = (\tilde{\sigma} - \sigma)(\tilde{\sigma} + \sigma)$$

differs from $(\tilde{\sigma} - \sigma)^2$ and cannot be matched by the quadratic loss for $\tilde{\sigma}$ with a different factor c.

We solve the problem of estimating σ^2 generally, by finding the function $c_n(R)$ for the factor c in $c\hat{\sigma}^2$ for penalty ratios $R > 0$ and sample sizes n. We derive first an identity that links the densities of χ^2. The χ^2 densities are related by the identity

$$y f_k(y) = \frac{1}{\Gamma_2(k)} \left(\tfrac{1}{2}y\right)^{k/2} \exp\left(-\tfrac{1}{2}y\right)$$
$$= k \, f_{k+2}(y), \tag{3.4}$$

derived directly from (3.2) using the reduction $2\,\Gamma_2(k+2)/\Gamma_2(k) = k$. The density of $c\hat{\sigma}^2$ is $b f_{n-1}(by)$, where $b = (n-1)/(c\sigma^2)$, so the expected loss due to overstatement, $\tilde{\sigma}^2 > \sigma^2$, is

$$
\begin{aligned}
Q_+ &= \frac{n-1}{c\sigma^2} \int_{\sigma^2}^{+\infty} \left(y - \sigma^2\right) f_{n-1}\left\{\frac{y(n-1)}{c\sigma^2}\right\} dy \\
&= \sigma^2 \int_{(n-1)/c}^{+\infty} \left(\frac{cz}{n-1} - 1\right) f_{n-1}(z)\, dz \\
&= \sigma^2 \left[c\left\{1 - F_{n+1}\left(\frac{n-1}{c}\right)\right\} - 1 + F_{n-1}\left(\frac{n-1}{c}\right) \right].
\end{aligned}
$$

Similar operations yield the expected loss due to understatement, $\tilde{\sigma}^2 < \sigma^2$,

$$
\begin{aligned}
Q_- &= \frac{(n-1)R}{c\sigma^2} \int_0^{\sigma^2} \left(\sigma^2 - y\right) f_{n-1}\left\{\frac{y(n-1)}{c\sigma^2}\right\} dy \\
&= R\sigma^2 \left\{ F_{n-1}\left(\frac{n-1}{c}\right) - c\, F_{n+1}\left(\frac{n-1}{c}\right) \right\},
\end{aligned}
$$

so the expected loss is

$$
Q = Q_+ + Q_- = \sigma^2 \left[c - 1 + (R+1)\left\{ F_{n-1}\left(\frac{n-1}{c}\right) - c\, F_{n+1}\left(\frac{n-1}{c}\right) \right\} \right].
$$

We search for the root of its derivative,

$$
\begin{aligned}
\frac{\partial Q}{\partial c} &= \sigma^2 \left\{ 1 - (R+1) F_{n+1}\left(\frac{n-1}{c}\right) \right\} \\
&\quad - \frac{(n-1)(R+1)\sigma^2}{c} \left\{ \frac{1}{c} f_{n-1}\left(\frac{n-1}{c}\right) - f_{n+1}\left(\frac{n-1}{c}\right) \right\}.
\end{aligned}
$$

The expression with the densities f_{n-1} and f_{n+1} in braces vanishes owing to (3.4). Therefore the optimal factor c is

$$
c_n(R) = \frac{n-1}{F_{n+1}^{-1}\left(\frac{1}{R+1}\right)}.
$$

The expected loss Q involves the variance σ^2 only as a factor, so $c_n(R)$ depends only on the sample size n and the penalty ratio R.

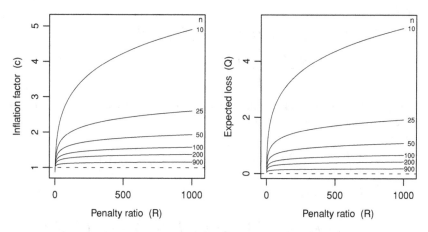

Fig. 3.2 The inflation factor $c_n(R)$ for estimating σ^2 with minimum piecewise linear expected loss, as a function of R and n, and the corresponding minimum expected loss $Q_n(R)$

The left-hand panel of Fig. 3.2 displays the solutions $c_n(R)$ for $R \in (5, 1000)$ and a selection of sample sizes n indicated at the right-hand margin. The right-hand panel plots the corresponding expected loss as a function of R for the same values of n and $\sigma^2 = 1$. The factor $c_n(R)$ increases with R from $(n-1)/(n+1)$, steeply for small R, and decreases with n, also steeply for small n. The minimum expected loss $Q_n(R)$ has similar features. The curves are much closer to linearity when the penalty ratio R is plotted on the log scale. Showing this is left for an exercise.

For estimating the standard deviation σ, we consider the estimators $\tilde{\sigma} = d\hat{\sigma}$, $d > 0$, and the piecewise quadratic loss. The distribution function of $\tilde{\sigma}$ is

$$P(\tilde{\sigma} < x) = P\left\{ \frac{(n-1)\hat{\sigma}^2}{\sigma^2} < vx^2 \right\}$$
$$= F_{n-1}\left(vx^2\right),$$

where $v = (n-1)/(d^2\sigma^2)$. The density of $\tilde{\sigma}$ is $2vx f_{n-1}(vx^2)$. The expected loss due to overstatement, $\tilde{\sigma} > \sigma$, is

$$Q_+ = \int_{\sigma}^{+\infty} 2vx(x-\sigma)^2 f_{n-1}\left(vx^2\right) dx$$
$$= \sigma^2 \int_{(n-1)/d^2}^{+\infty} \left(\frac{d\sqrt{z}}{\sqrt{n-1}} - 1 \right)^2 f_{n-1}(z) dz,$$

after applying the transformation $z = vx^2$. We expand the square in the integrand and express each term as a scalar multiple of a density, which is then easy to integrate. In addition to (3.4), we use the identity

$$\sqrt{z}\, f_k(z) = \frac{1}{\sqrt{2}\, \Gamma_2(k)} \left(\tfrac{1}{2}z\right)^{k/2-1/2} \exp\left(-\tfrac{1}{2}z\right)$$

$$= \frac{\sqrt{2}\, \Gamma_2(k+1)}{\Gamma_2(k)}\, f_{k+1}(z) = \sqrt{k}\, H_{k+1} f_{k+1}(z), \qquad (3.5)$$

with the constant H_n introduced after (3.3), where a similar identity was used. The expected loss due to overstatement is

$$Q_+ = \sigma^2 \int_{(n-1)/d^2}^{+\infty} \left\{ d^2 f_{n+1}(z) - 2d\, H_n\, f_n(z) + f_{n-1}(z) \right\} dz$$

$$= \sigma^2 \left(d^2 - 2d\, H_n + 1 \right)$$

$$- \sigma^2 \left\{ d^2 F_{n+1}\left(\frac{n-1}{d^2}\right) - 2d\, H_n\, F_n\left(\frac{n-1}{d^2}\right) + F_{n-1}\left(\frac{n-1}{d^2}\right) \right\}.$$

For Q_- we have a similar expression,

$$Q_- = R\sigma^2 \int_0^{(n-1)/d^2} \left(\frac{d\sqrt{z}}{\sqrt{n-1}} - 1 \right)^2 f_{n-1}(z)\, dz$$

$$= R\sigma^2 \left\{ d^2 F_{n+1}\left(\frac{n-1}{d^2}\right) - 2d\, H_n\, F_n\left(\frac{n-1}{d^2}\right) + F_{n-1}\left(\frac{n-1}{d^2}\right) \right\},$$

and so the expected loss is

$$Q = \sigma^2 \Big[d^2 - 2d\, H_n + 1 + (R-1)$$

$$\times \left\{ d^2 F_{n+1}\left(\frac{n-1}{d^2}\right) - 2d\, H_n\, F_n\left(\frac{n-1}{d^2}\right) + F_{n-1}\left(\frac{n-1}{d^2}\right) \right\} \Big].$$

The minimum of this function of d is found by the Newton-Raphson algorithm. The derivative of Q is

$$s = 2\sigma^2 \left[d - H_n + (R-1) \left\{ d\, F_{n+1}\left(\frac{n-1}{d^2}\right) - H_n\, F_n\left(\frac{n-1}{d^2}\right) \right\} \right],$$

after the terms involving densities f cancel out owing to (3.4) and (3.5). The root of s, denoted by $d_n(R)$, does not depend on σ^2. Figure 3.3 summarises this function for the same sets of values of n and R and with the same layout as Fig. 3.2. Figure 3.4 offers a more direct comparison of the optimal inflation factors c and d. Intuition suggests that c should be close to d^2. However, the ratio c/d^2 is for all R smaller than 1.0, by a wide margin for small n. The minimum loss for estimating σ with piecewise quadratic loss is much greater than the minimum loss for estimating σ^2 with piecewise linear loss. This does not mean that estimation of one quantity should

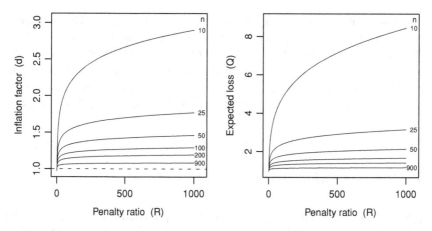

Fig. 3.3 The optimal factor $d_n(R)$ for estimating σ with piecewise quadratic loss and the corresponding minimum loss $Q_n(R)$

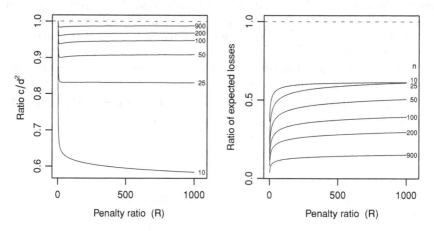

Fig. 3.4 Comparison of the inflation factors $c_n(R)$ and $d_n(R)$ from Figs. 3.2 and 3.3

be preferred over the other, merely that the perspectives and priorities implied by the two estimation problems should not be confused.

3.3 Variance Versus a Constant

This section addresses the problem of comparing the variance of a normal random sample with a constant. We review how this problem is treated by hypothesis testing, and then find a solution with utilities. Throughout, we assume that there is a positive critical value σ_0^2 of the variance σ^2; one course of action is appropriate when $\sigma^2 < \sigma_0^2$, and another when $\sigma^2 > \sigma_0^2$.

Suppose we have a random sample $X_1, \ldots, X_n$ from $\mathcal{N}(\mu, \sigma^2)$. The parameters μ and σ^2 are not known. We choose $\hat{r} = (n-1)\hat{\sigma}^2/\sigma_0^2$ as the test statistic. This statistic has χ_{n-1}^2 distribution if $\sigma^2 = \sigma_0^2$. For a test of size 0.05, we set the critical region to $(C_U, +\infty)$, where C_U is the 95th percentile of χ_{n-1}^2 distribution. In general, $P(\hat{r}\sigma_0^2/\sigma^2 > C_U) = 0.05$. Therefore, if $\sigma^2 < \sigma_0^2$, $P(\hat{r} > C_U) < 0.05$. Thus, the test is proper. The power of the test when $\sigma^2 > \sigma_0^2$ is $P(\hat{r} > C_U) > 0.05$, so the test is unbiased.

3.3.1 Decision with Utilities

Suppose small variance, $\sigma^2 < \sigma_0^2$, corresponds to a preferred state of affairs, but the penalty for an unjustified claim that the variance is small, is harsh. To set the problem so that it resembles the scenario of testing a hypothesis, we assume a piecewise constant loss. Suppose the decision will be based on the sign of $\rho\hat{\sigma}^2 - \sigma_0^2$, with a suitable positive constant (factor) $\rho > 0$. If the claim of small variance is made falsely, when $\rho\hat{\sigma}^2 < \sigma_0^2 < \sigma^2$, we incur loss R. If the claim is not made, but should have been made, when $\sigma^2 < \sigma_0^2 < \rho\hat{\sigma}^2$, the loss is one unit. When $\rho\hat{\sigma}^2$ and σ^2 are both greater or both smaller than σ_0^2, no loss is incurred because the appropriate course of action is selected, even when $\rho\hat{\sigma}^2 \neq \sigma_0^2$.

Earlier we established that $(n-1)\hat{\sigma}^2/\sigma^2$ has χ_{n-1}^2 distribution. Let U be a variable with this distribution. We represent σ^2 by the random variable

$$\frac{(n-1)\hat{\sigma}^2}{U},$$

which has the scaled inverse gamma distribution with density

$$\frac{(n-1)\hat{\sigma}^2}{u^2} f_{n-1}\left\{\frac{(n-1)\hat{\sigma}^2}{u}\right\}.$$

We have converted an unknown constant, σ^2, into a random variable, to represent our uncertainty about its value, and the sample quantity $\hat{\sigma}^2$, originally a random variable, is now regarded as a constant. We made a similar conversion in Sect. 2.5 (page 29).
When $\rho\hat{\sigma}^2 < \sigma_0^2$, the expected loss is

$$Q_+ = R \int_{\sigma_0^2}^{+\infty} \frac{(n-1)\hat{\sigma}^2}{u^2} f_{n-1}\left\{\frac{(n-1)\hat{\sigma}^2}{u}\right\} du \,;$$

otherwise it is

$$Q_- = \int_0^{\sigma_0^2} \frac{(n-1)\hat{\sigma}^2}{u^2} f_{n-1}\left\{\frac{(n-1)\hat{\sigma}^2}{u}\right\} du.$$

After the transformation that simplifies the argument of f_{n-1}, these expected losses become

$$Q_+ = R \int_0^\tau f_{n-1}(z) \, dz = R F_{n-1}(\tau)$$

$$Q_- = \int_\tau^{+\infty} f_{n-1}(z) \, dz = 1 - F_{n-1}(\tau), \qquad (3.6)$$

where $\tau = (n-1)\hat{\sigma}^2/\sigma_0^2$. For very small ρ, Q_+ applies. As we increase ρ, we reach a point at which the expected loss switches from Q_+ to Q_-. The value of the ratio $\rho = \sigma_0^2/\hat{\sigma}^2$ for which $Q_+ = Q_-$ is the equilibrium, and is denoted by ρ_R. If we set ρ to ρ_R, then we switch from one decision to the other at τ, where $Q_+ = Q_-$. Since Q_+ is an increasing function of the ratio $\hat{\sigma}^2/\sigma_0^2$, and Q_- a decreasing function, $\max(Q_+, Q_-)$ would be greater for any other choice of ρ. When $\rho_R \hat{\sigma}^2 = \sigma_0^2$, the choice between the two courses of action is immaterial, because the expected losses are identical. When $\rho_R \hat{\sigma}^2 < \sigma_0^2$ we claim that the variance is small, and refrain from the claim otherwise. The expected losses in (3.6) are equal when

$$\tau = F_{n-1}^{-1}\left(\frac{1}{R+1}\right),$$

that is, for the $1/(R+1)$-quantile of the χ_{n-1}^2 distribution. The corresponding factor ρ is $\rho_R = (n-1)/\tau$, so our estimator of σ^2 is

$$\rho_R \hat{\sigma}^2 = \frac{(n-1)\hat{\sigma}^2}{F_{n-1}^{-1}\left(\frac{1}{R+1}\right)}.$$

With the piecewise constant loss, we would get the same solution, that is, $\hat{\sigma}\sqrt{\rho_R}$, if we had to decide whether to make the claim based on $\hat{\sigma}$ instead of on $\hat{\sigma}^2$. Of course, with piecewise linear and quadratic loss functions we obtain different solutions. This is undesirable; the scale on which we operate should make no difference. It does make some if we fail to transform the loss function in accordance with the transformation of the target.

3.3.2 Multiplicative Loss

The decision rule we adopted can be formulated as comparing $\hat{\sigma}^2/\sigma_0^2$ against a constant. This motivates the loss function defined as

$$L_+ = R\left(\frac{\sigma^2}{\sigma_0^2} - 1\right)$$

$$L_- = \frac{\sigma_0^2}{\sigma^2} - 1, \tag{3.7}$$

when we incorrectly decide that $\sigma^2 < \sigma_0^2$ and $\sigma^2 > \sigma_0^2$, respectively. We base the decision on $\rho\hat{\sigma}^2$, and seek the equilibrium value of ρ or of $\rho\hat{\sigma}^2/\sigma_0^2$, for which the balance function $Q_+ - Q_- = \mathrm{E}(L_+) - \mathrm{E}(L_-)$ vanishes.

The two parts of the expected loss are obtained by integrating the loss functions in (3.7):

$$
\begin{aligned}
Q_+ &= R \int_{\sigma_0^2}^{+\infty} \left(\frac{u}{\sigma_0^2} - 1 \right) \frac{(n-1)\hat{\sigma}^2}{u^2} f_{n-1}\left\{ \frac{(n-1)\hat{\sigma}^2}{u} \right\} du \\
&= R \int_0^\tau \left(\frac{\tau}{z} - 1 \right) f_{n-1}(z)\,dz \\
&= \frac{R\tau}{n-3} F_{n-3}(\tau) - R F_{n-1}(\tau) \\
Q_- &= \int_0^{\sigma_0^2} \left(\frac{\sigma_0^2}{u} - 1 \right) \frac{(n-1)\hat{\sigma}^2}{u^2} f_{n-1}\left\{ \frac{(n-1)\hat{\sigma}^2}{u} \right\} du \\
&= \int_\tau^{+\infty} \left(\frac{z}{\tau} - 1 \right) f_{n-1}(z)\,dz \\
&= \frac{n-1}{\tau} \{1 - F_{n+1}(\tau)\} - 1 + F_{n-1}(\tau), \tag{3.8}
\end{aligned}
$$

where $\tau = (n-1)\hat{\sigma}^2/\sigma_0^2$, as defined after (3.6). The concluding identities for Q_+ and Q_- in (3.8) are obtained by applying (3.4) with $k = n - 3$ and $k = n - 1$, respectively. The expected losses Q_+ and Q_- depend on the variances $\hat{\sigma}^2$ and σ_0^2 only through τ. The derivative of the balance function $\Delta Q = Q_+ - Q_-$ is

$$\frac{\partial \Delta Q}{\partial \tau} = \frac{R}{n-3} F_{n-3}(\tau) + \frac{n-1}{\tau^2} \{1 - F_{n+1}(\tau)\},$$

after all the terms involving densities f cancel out, owing to (3.4). The derivative is positive, so ΔQ has a unique root. From the solution τ_R we obtain the equilibrium ratio $\rho_R = \tau_R/(n-1)$ and the estimator $\hat{\sigma}^2 \tau_R/(n-1)$.

Figure 3.5 presents the solutions for $R \in (1, 1000)$ and a selection of sample sizes n on the linear and log scales for R. For fixed R, the equilibrium ratio decreases with n and converges to 1.0 as $n \to +\infty$. For fixed sample size n, $\rho_R(n)$ increases with R. In brief, when we are averse to false declarations of small variance, we inflate the estimate $\hat{\sigma}^2$, to a greater extent with small samples (a lot of uncertainty) and large penalty ratios (greater aversion), steeply at first, and then progressively more gradually. On the log scale, the functions $\rho_R(n)$ are close to linearity, especially for $R \gg 1$. The multiplicative scale is much better suited for R.

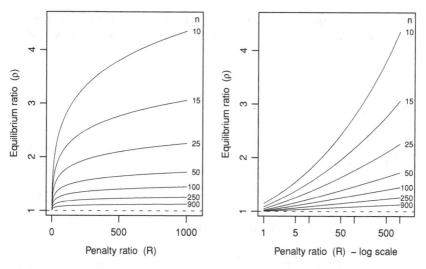

Fig. 3.5 The equilibrium ratios for comparing a variance to a constant with multiplicative loss

3.4 Estimating the Variance Ratio

Suppose we have two independent random samples of respective sizes n_1 and n_2 from distinct normal distributions and we want to estimate the ratio of their variances $r = \sigma_1^2/\sigma_2^2$. The established estimator of r is the ratio of the usual unbiased estimators, $\hat{r} = \hat{\sigma}_1^2/\hat{\sigma}_2^2$. It is neither unbiased nor efficient; its principal attraction is the reference to the F distribution. The scaled version of the ratio, $\hat{r}\sigma_2^2/\sigma_1^2$, has F distribution with $k_1 = n_1 - 1$ and $k_2 = n_2 - 1$ degrees of freedom. The F distribution has the density

$$g(y; k_1, k_2) = \frac{\Gamma_2(k_1 + k_2)}{\Gamma_2(k_1)\,\Gamma_2(k_2)} \left(\frac{k_1}{k_2}\right)^{k_1/2} y^{k_1/2-1} \left(1 + \frac{k_1 y}{k_2}\right)^{-(k_1+k_2)/2}. \qquad (3.9)$$

Denote by G the corresponding distribution function. The expectation of the distribution is $k_2/(k_2 - 2)$ when $k_2 > 2$; otherwise the expectation is not defined. The variance of the distribution, defined for $k_2 > 4$, is

$$\frac{2k_2^2(k_1 + k_2 - 2)}{k_1(k_2 - 2)^2(k_2 - 4)}.$$

For $k_2 > 2$, the estimator $\hat{r}(k_2 - 2)/k_2$ is unbiased, and the estimator $c\hat{r}$ attains minimum MSE for

$$c^* = \frac{k_1(k_2 - 2)(k_2 - 4)}{k_2(k_1 k_2 - 2k_1 + 2k_2 - 2)}, \qquad (3.10)$$

assuming that $k_2 > 4$.

We evaluate the expected loss of estimators $c\hat{r}$ with respect to the piecewise linear loss; the loss is $R(\hat{r} - r)$ when $\hat{r} > r$ and it is $r - \hat{r}$ when $\hat{r} < r$. The density of $c\hat{r}$ is

$$\frac{1}{cr} g\left(\frac{y}{cr}; k_1, k_2\right)$$

and the expected loss is

$$Q = \frac{1}{cr}\left\{R\int_r^{+\infty}(y-r)g\left(\frac{y}{cr}; k_1, k_2\right)dy + \int_0^r(r-y)g\left(\frac{y}{cr}; k_1, k_2\right)dy\right\}$$

$$= r\left\{R\int_{1/c}^{+\infty}(cz-1)g(z; k_1, k_2)dz + \int_0^{1/c}(1-cz)g(z; k_1, k_2)dz\right\}.$$

To avoid any numerical integration, we use the identity

$$x\,g(x; k_1, k_2) = \frac{k_1}{k_1+2}\,g(h_2\,x; k_1+2, k_2-2),\tag{3.11}$$

where

$$h_2 = \frac{k_1(k_2-2)}{k_2(k_1+2)}.$$

The identity is proved directly from the definition of g in (3.9). With this result, we have

$$Q = \frac{k_1 cr\,R}{k_1+2}\int_{1/c}^{+\infty}g(h_2 z; k_1+2, k_2-2)\,dz$$

$$-\frac{k_1 cr}{k_1+2}\int_0^{1/c}g(h_2 z; k_1+2, k_2-2)\,dz - r\left\{R - (R+1)G\left(\frac{1}{c}; k_1, k_2\right)\right\}$$

$$= \frac{k_2\,cr}{k_2-2}\left\{R - (R+1)G\left(\frac{h_2}{c}; k_1+2, k_2-2\right)\right\}$$

$$-r\left\{R - (R+1)G\left(\frac{1}{c}; k_1, k_2\right)\right\}.$$

We find the minimum of this function of c as the root of its derivative,

$$\frac{\partial Q}{\partial c} = \frac{k_2\,r}{k_2-2}\left\{R - (R+1)G\left(\frac{h_2}{c}; k_1+2, k_2-2\right)\right\}$$

$$-\frac{r(R+1)}{c}\left\{\frac{1}{c}f\left(\frac{1}{c}; k_1, k_2\right) - \frac{k_1}{k_1+2}f\left(\frac{h_2}{c}; k_1+2, k_2-2\right)\right\}.$$

The expression in the concluding line vanishes owing to (3.11). Hence, the (unique) root of the derivative is

$$c^* = \frac{h_2}{G^{-1}\left(\frac{R}{R+1}; k_1 + 2, k_2 - 2\right)}.$$

This is the sole minimum of Q, because

$$\frac{\partial^2 Q}{\partial c^2} = \frac{h_2 k_2 r(R+1)}{c^2(k_2 - 2)} g\left(\frac{h_2}{c}; k_1 + 2, k_2 - 2\right) = \frac{r(R+1)}{c^3} g\left(\frac{1}{c}; k_1, k_2\right)$$

is positive. The minimum attained is

$$Q^* = r(R+1) G\left\{\frac{1}{h_2} G^{-1}\left(\frac{R}{R+1}; k_1 + 2, k_2 - 2\right); k_1, k_2\right\} - rR. \qquad (3.12)$$

We do not have a simple comparison of $Q(c^*)$ with the expected loss $Q(1)$. The derivative $\partial Q/\partial c$ is a decreasing function of R; therefore the optimum factor c^* is an increasing function of R.

Instead of the variance ratio r we may estimate the ratio of the standard deviations $\sqrt{r}$. We explore the estimators $d\sqrt{\hat{r}}$ for $d > 0$. The expected (piecewise linear) loss is

$$Q = \sqrt{r}\left\{R\int_{1/d^2}^{+\infty}(d\sqrt{z} - 1) g(z)\,dz + \int_0^{1/d^2}(1 - d\sqrt{z}) g(z)\,dz\right\}.$$

We use the following identity, similar to (3.11):

$$\sqrt{x}\,g\,(x; k_1, k_2) = \kappa(k_1, k_2)\, g(h_1 x; k_1 + 1, k_2 - 1), \qquad (3.13)$$

where

$$h_1 = \frac{k_1}{k_1 + 1}\frac{k_2 - 1}{k_2}$$

$$\kappa(k_1, k_2) = \frac{\Gamma_2(k_1 + 1)\,\Gamma_2(k_2 - 1)}{\Gamma_2(k_1)\,\Gamma_2(k_2)}\frac{k_2 - 1}{k_1 + 1}\frac{\sqrt{k_1}}{\sqrt{k_2}}.$$

The identity in (3.13) is proved directly from (3.9). With it, we have

$$Q = \sqrt{r}\,\frac{d\,\kappa(k_1, k_2)}{h_1}\left\{R - (R+1) G\left(\frac{h_1}{d^2}; k_1 + 1, k_2 - 1\right)\right\}$$
$$- \sqrt{r}\left\{R - (R+1) G\left(\frac{1}{d^2}; k_1, k_2\right)\right\},$$

an expression similar to its counterpart for $c\hat{r}$ in (3.12). The derivatives of this expected loss with respect to d are

$$\frac{\partial Q}{\partial d} = \frac{\sqrt{r}\,\kappa(k_1, k_2)}{h_1}\left\{R - (R+1)\,G\!\left(\frac{h_1}{d^2}; k_1 - 1, k_2 + 1\right)\right\}$$

$$\frac{\partial^2 Q}{\partial d^2} = \frac{2\sqrt{r}\,(R+1)\,\kappa(k_1, k_2)}{d^3}\,g\!\left(\frac{h_1}{d^2}; k_1 - 1, k_2 + 1\right),$$

so Q has a unique minimum at

$$d^* = \sqrt{\frac{h_1}{G^{-1}\!\left(\frac{R}{R+1}; k_1 - 1, k_2 + 1\right)}}.$$

Note that d^{*2} and c^* differ only in the degrees of freedom in the argument of the quantile G^{-1} and in the constant in the numerator. Of course, the minimum expected loss estimators of r and $\sqrt{r}$ differ, because the piecewise linear loss functions in the two cases represent different losses.

3.5 Problems, Exercises and Suggested Reading

1. Explore why piecewise constant loss function for estimating a variance makes no sense. Generalise your conclusion to other targets of estimation.
2. The relative MSE of one estimator, $\hat{\theta}_1$, against another, $\hat{\theta}_2$, when both of them have the same target θ, is defined as $\mathrm{MSE}(\hat{\theta}_1; \theta)/\mathrm{MSE}(\hat{\theta}_2; \theta)$. The relative root-MSE is defined similarly. Reproduce the right-hand panel of Fig. 3.1 with the relative root-MSEs of the efficient and unbiased estimators of σ related to the root-MSE of the naive estimator. Explain why this plot has a much better resolution.
3. Prove the identities in (3.11) and (3.13). Find some other classes of distributions for which similar identities apply.
4. Verify the expressions for the expectation and variance of the F distribution and that the estimators listed in the text are unbiased and efficient in the class $\{c\hat{r}\}$. Show that c^* in (3.10) is smaller than unity. (In fact, it is smaller than $1 - 4/k_2$.) Confirm your (analytical) conclusion by a suitable graph.
5. Find the unbiased and efficient estimators of $\sqrt{r}$ in the class $d\sqrt{\hat{r}}$ and compare the coefficients d (or d^2) with their counterparts for estimating r. Compare graphically the optimal factors c^* and d^{*2}.
6. Find the estimator of the variance ratio with the piecewise quadratic loss.
7. Derive the loss function for σ that matches piecewise linear loss for σ^2.
8. Explore the connection of the piecewise linear loss with estimating quantiles. Suggested reading on quantile regression: Koenker (2005).
9. Explore the analytical difficulties with piecewise absolute, linear and quadratic loss functions specified for estimating $\log(\sigma^2)$ or $\log(1 + \sigma^2)$ or comparing them to a constant.

10. Plot the densities of χ^2 distributions, suitably scaled, and compare them with the normal distribution that has the same mean and variance. Decide from which number of degrees of freedom onwards the χ^2 is indistinguishable from the normal. A more ambitious exercise: do the same with the densities of the F distribution. Suggested reading for background on distributions derived from the normal, which include F and χ^2: Simon (2004).

11. A commonly adopted standard for inference is to estimate a parameter efficiently and to estimate the MSE (sampling variance if the estimator is unbiased) of this estimator without bias. (Why not estimate the root-MSE or standard error without bias?) Do you agree with this wholeheartedly? See Longford (2013) for an alternative view.

12. Reading of historical interest: Markowitz (1968) and Stuart (1969).

13. Hypotheses about variances are often formulated in the context of variance component models (McCulloch et al., 2006). An application that provided a powerful stimulus for their development in animal breeding is Henderson (1984a). For a review of criteria for variance estimation, see Henderson (1984b).

14. Monographs on decision theory: French and Insua (2004), Liese and Miescke (2008) and Rapoport (2010, Part I); their precursor is Le Cam (1986).

References

French, S., & Insua, D. R. (2004). *Statistical decision theory*. New York: Oxford University Press.

Henderson, C. R. (1984a). *Applications of linear models in animal breeding*. Guelph: University of Guelph.

Henderson, C. R. (1984b). ANOVA, MIVQUE, REML, and ML algorithms for estimation of variances and covariances. In H. A. David & H. T. David (Eds.), *Statistics: An appraisal* (pp. 257–280). Iowa State University.

Koenker, R. (2005). *Quantile regression*. Cambridge: Cambridge University Press.

Le Cam, L. (1986). *Asymptotic methods in statistical decision theory*. New York: Springer-Verlag.

Liese, F., & Miescke, K. J. (2008). *Statistical decision theory. Estimation, testing and selection*. Springer-Verlag, New York.

Longford, N. T. (2013). Assessment of precision with aversity to overstatement. *South African Statistical Journal, 47*, 49–59.

Markowitz, E. (1968). Minimum mean-square-error of estimation of the standard deviation of the normal distribution. *The American Statistician, 22*, 26.

McCulloch, C. E., Searle, S. R., & Casella, G. (2006). *Variance components* (2nd ed.). New York: Wiley.

Rapoport, A. (2010). *Decision theory and decision behaviour*. Dordrecht: Kluwer.

Simon, M. K. (2004). *Probability distributions involving Gaussian random variables*. New York: Springer-Verlag.

Stuart, A. (1969). Reduced mean-square-error estimation of σ^p in normal samples. *The American Statistician, 23*, 27–28.

Chapter 4
The Bayesian Paradigm

The Bayesian paradigm is founded on a probabilistic way of updating the prior information about the inferential targets (population quantities) by the information contained in the collected data. A model is posited for the collected data, in essentially the same way as in the frequentist perspective, and prior information is expressed as a joint distribution for the model parameters. This distribution is independent of the data. The first-stage outcome of a Bayesian analysis is the posterior distribution of the model parameters, which then requires further processing.

4.1 The Bayes Theorem

Suppose outcomes $\mathbf{x} = (x_1, \ldots, x_n)$ have been generated as a random draw from a distribution with joint density $f(\mathbf{y} \mid \boldsymbol{\theta})$, where $\boldsymbol{\theta}$ is a vector of parameters and $\mathbf{y}$ is the n-variate argument of f. The value of $\boldsymbol{\theta}$ is known to belong to a set $\boldsymbol{\Theta}$, called the parameter space. The collection of these distributions, $\{f(\mathbf{y} \mid \boldsymbol{\theta}); \boldsymbol{\theta} \in \boldsymbol{\Theta}\}$, is referred to as the *model*. The task is to make an inferential statement about the vector $\boldsymbol{\theta}$, one of its subvectors, or a transformation $\zeta(\boldsymbol{\theta})$. We write the model density f as conditional on $\boldsymbol{\theta}$, because $\boldsymbol{\theta}$ is treated as a random vector. The information about $\boldsymbol{\theta}$ that is available independently of the observed sample is summarised by a prior distribution, with density $\pi(\boldsymbol{\theta})$. The posterior distribution of $\boldsymbol{\theta}$ is defined as its conditional distribution given the outcomes $\mathbf{x}$. Its density is denoted by $g(\boldsymbol{\theta} \mid \mathbf{x})$.

The Bayes theorem for absolutely continuous densities f and π states that

$$g(\boldsymbol{\theta} \mid \mathbf{x}) = \frac{f(\mathbf{x} \mid \boldsymbol{\theta})\pi(\boldsymbol{\theta})}{\int \ldots \int f(\mathbf{x} \mid \boldsymbol{\xi})\pi(\boldsymbol{\xi})\,\mathrm{d}\boldsymbol{\xi}}. \tag{4.1}$$

It can be interpreted as switching the roles of $\mathbf{x}$ and $\boldsymbol{\theta}$ from the model, in which we specify the range of behaviours of potential samples $\mathbf{x}$ given $\boldsymbol{\theta}$, to the posterior, in which we update our (prior) information about $\boldsymbol{\theta}$ by the recorded data $\mathbf{x}$.

N. T. Longford, *Statistical Decision Theory*,
SpringerBriefs in Statistics, DOI: 10.1007/978-3-642-40433-7_4,
© The Author(s) 2013

For discrete data, we replace the density f by a probability function. The prior distribution is in most cases absolutely continuous. When it is discrete, its probabilities are substituted for π and the integral(s) in the denominator are replaced by summation(s).

The posterior density $g(\theta \,|\, \mathbf{x})$ is the sole outcome of the analysis thus far and any inferential statements are based on it. The posterior expectation

$$\hat{\theta} = \mathrm{E}\,(\theta \,|\, \mathbf{x}) = \int \cdots \int \xi \, g(\xi \,|\, \mathbf{x})\,\mathrm{d}\xi$$

can be adopted as an estimate of θ, and the posterior variance (matrix)

$$\mathrm{var}\,(\theta \,|\, \mathbf{x}) = \int \cdots \int \left(\xi - \hat{\theta}\right)\left(\xi - \hat{\theta}\right)^{\top} g(\xi \,|\, \mathbf{x})\,\mathrm{d}\xi$$

can be regarded as a measure of uncertainty about θ. For univariate θ, small posterior variance is desirable and the tails of the posterior distribution are studied to understand how plausible certain extreme values of θ are.

Evaluation of the integral in (4.1) would cause a problem in many settings, but it can be sidestepped by replacing it with the constant C for which $Cf(\mathbf{x} \,|\, \theta)\pi(\theta)$ is a proper density, that is, for which its integral is equal to unity. The constant can often be guessed by matching the product $f(\mathbf{x} \,|\, \theta)\pi(\theta)$ to a known density.

Example

Suppose $\mathbf{x} = (x_1, \ldots, x_n)^{\top}$ is a (realised) random sample from $\mathcal{N}(\mu, 1)$. Denote its sample mean by $\bar{x}$. Suppose the prior distribution for μ is also normal, $\mathcal{N}(\delta, 1/n_0)$, where δ and $n_0 > 0$ are given. In this setting, the model and prior densities are

$$f(\mathbf{x} \,|\, \mu) = \frac{1}{(2\pi)^{n/2}} \exp\left\{-\frac{1}{2}\sum_{i=1}^{n}(x_i - \mu)^2\right\}$$

$$\pi(\mu) = \frac{\sqrt{n_0}}{\sqrt{2\pi}} \exp\left\{-\frac{n_0(\mu - \delta)^2}{2}\right\},$$

respectively. Their product, the numerator in (4.1), is

$$f(\mathbf{x} \,|\, \mu)\,\pi(\mu) = \frac{\sqrt{n_0}}{(2\pi)^{n/2\,+\,1/2}} \exp\left[-\frac{1}{2}\left\{n_0(\mu - \delta)^2 + \sum_{i=1}^{n}(x_i - \mu)^2\right\}\right]$$

$$= \frac{\sqrt{n_0 + n}}{\sqrt{2\pi}} \exp\left\{-\frac{(n_0 + n)}{2}\left(\mu - \tilde{\delta}\right)^2\right\}$$

$$\times \frac{\sqrt{n_0}}{\sqrt{(2\pi)^n (n_0 + n)}} \exp\left\{\frac{(n_0 + n)\tilde{\delta}^2 - n_0\delta^2 - \sum_{i=1}^{n} x_i^2}{2}\right\}, \tag{4.2}$$

where

$$\tilde{\delta} = \frac{n_0\delta + n\bar{x}}{n_0 + n}. \tag{4.3}$$

The second line in (4.2) is the density of $\mathcal{N}\{\tilde{\delta}, 1/(n_0 + n)\}$ and the concluding line does not depend on μ. It has to be equal to the reciprocal of the denominator in the Bayes theorem, (4.1); otherwise $g(\mu \mid \mathbf{x})$ would not be a density. Hence the posterior distribution of μ is $\mathcal{N}\{\tilde{\delta}, 1/(n_0 + n)\}$.

It is very convenient that the posterior distribution of μ is normal, just like the observations x_i, $i = 1, \ldots, n$, their summary $\bar{x}$ and the prior π. A prior for which the posterior belongs to the same class of distributions is said to be conjugate for the data distribution (the model). Conjugate priors are preferred in Bayesian analysis because they lead to posteriors that are easy to handle. We note that this is rarely relevant to the client's perspective, although the analyst may plead that posterior distributions are intractable for some classes of priors.

The posterior distribution $\mathcal{N}\{\tilde{\delta}, 1/(n_0 + n)\}$ coincides with the (frequentist sampling) distribution of the sample mean based on $n_0 + n$ observations, n_0 of which are additional to the realised data. The additional observations, which we might call prior data, are a random sample from $\mathcal{N}(\delta, 1)$, independent of the n realised (genuine) observations.

The claim, often made in the literature, that Bayesian analysis is essential for incorporating prior information is poorly founded in this example, because the prior information could be represented in a frequentist analysis by a set of fictitious (prior) observations. Admittedly, these observations are not real, but the nature of the prior distribution is not much different. Both have to be obtained from the client by elicitation. The idea of prior (hypothetical) observations may appeal to the client much better than the often less familiar concept of a (prior) distribution.

As an estimator, the posterior expectation in (4.3) can be interpreted as a compromise between two alternatives: the data-based estimator $\bar{x}$ and the prior estimator δ, with respective variances $1/n_0$ and $1/n$. Selecting either of them is patently inferior to their convex linear combination (composition).

For a loss function L, the posterior expected loss is defined as the expected loss with respect to the posterior distribution:

$$Q_{\mathrm{D}} = \int \cdots \int L(\mathrm{D}, \boldsymbol{\theta}) \, g(\boldsymbol{\theta} \mid \mathbf{x}) \, \mathrm{d}\boldsymbol{\theta},$$

where D is the contemplated action. In practice, we evaluate Q_{A} and Q_{B} for the alternative courses of action A and B, and choose the one with smaller value of Q.

In many instances, including the example of estimating the normal mean, the posterior density depends on the data only through one or a few summaries. Such a set of summaries is called a set of *sufficient statistics* (for one or a set of parameters).

For example, $\bar{x}$ is a sufficient statistic for μ. In some problems in which there is a single sufficient statistic, such as $t(\mathbf{x}) = \bar{x}$ for μ, we can anticipate that the solution will have the form of preferring A when $t(\mathbf{x}) < T$ and preferring B otherwise. The value of T may depend on the sample size and other quantities that characterise the data collection or generation process. We can solve such problems in advance of data collection by finding the threshold value T. When the data is collected, in a dataset $\mathbf{x}$, the decision is made immediately by evaluating the statistic $t(\mathbf{x})$ and comparing it with T. Apart from not holding up the client in his business, having a decision rule in advance of the study is invaluable for design, for planning how the study should be organised. This is dealt with in detail in Chap. 8.

The established way of conducting a Bayesian analysis centers on deriving the posterior distribution of the relevant parameter for the collected data and the elicited or otherwise declared prior information. In the final step, the expected losses of the alternative courses of action are compared. We regard the prior information to be as important as the recorded data and ascribe to it comparable status and importance. In this approach, we discard default synthetic choices, such as non-informative priors, which are suitable in classroom exercises in which the expert is distant and unreachable. Elicitation of the prior is a complex and often time-consuming process of discussing the background to the client's problem and the analyst's mode of operation, and in a technically (computationally) oriented curriculum it naturally does not figure prominently. In practice, it is often a key ingredient of a salient analysis.

A cautious and responsible expert is unlikely to provide a single prior distribution. He may be concerned that a prior similar to the one put forward may equally well have been selected, but it would lead to a different conclusion. A more constructive method of elicitation of the prior concludes with a plausible set of prior distributions or their parameters. Such a set may be constructed by gradual reduction from a much larger set until the expert is no longer willing to concede any further reduction, being concerned that a prior might be inappropriately ruled out. When the expert is not available for elicitation the analyst will retain greater integrity by declaring a (wide) set of plausible priors that reflect his understanding of the (distant) expert's perspective, or what he anticipates the expert might declare, and may contact the expert with the plausible results (decisions) after concluding the analysis. In brief, requiring a single prior is an unnecessary straitjacket.

Further, we invert the process of generating the posterior for each prior. We split the set of priors into subsets that correspond to the preference for each course of action. The priors for which the values of the expected posterior loss for the two actions coincide are called *equilibrium*. If all the plausible priors belong to one subset of priors, we have an unequivocal conclusion, the choice of the action that would be preferred with any plausible prior. Otherwise we face an *impasse* (no clear decision), which can be resolved by reviewing the set of plausible priors and reducing it. In this approach, the set of plausible priors is declared in advance. As an alternative, the expert may be presented with the description of the two subsets, or of the set of equilibrium priors that divide them, and asked whether one of the subsets could be ruled out entirely.

We do not insist on declaring a single loss function either. Instead, we work with a set of plausible loss functions and, in effect, solve the problem for every one of these functions. In most settings, it suffices to solve the problem for the (extreme) plausible loss functions that delimit their range. Of course, the more inclusive (liberal) we are in specifying the prior information (sets of loss functions and prior distributions), the more likely we are to conclude the analysis with an impasse, so detailed elicitation may be invaluable.

4.2 Comparing Two Normal Random Samples

We address the problem of deciding whether the expectation of one normal random sample is greater than the expectation of another. This is one of the basic problems in statistics that is commonly resolved by hypothesis testing; see Sect. 2.5. Here we define several classes of loss functions and find which sign of the difference of the means is associated with smaller expected loss.

Suppose we have independent random samples from normal distributions $\mathcal{N}(\mu_1, \sigma^2)$ and $\mathcal{N}(\mu_2, \sigma^2)$, with respective sizes n_1 and n_2. Let $m = 1/n_1 + 1/n_2$. The two samples have identical variances, equal to σ^2, which we assume to be known. No generality is lost by assuming that $\sigma^2 = 1$, because instead of observations x we could work with x/σ. Denote by $\widehat{\Delta}$ the difference of the sample means; $\widehat{\Delta} = \bar{x}_1 - \bar{x}_2$. In the frequentist perspective, it is unbiased for $\Delta = \mu_1 - \mu_2$ and its sampling variance is $m\sigma^2$. For Δ we have the prior $\mathcal{N}(\delta, q\sigma^2)$; δ and $q > 0$ are given, although later we deal with the problem when we only have ranges (intervals) of plausible values for them.

As in Chap. 2, we define the piecewise quadratic loss function as $L_- = \Delta^2$ when we choose the negative sign but $\Delta > 0$, and as $L_+ = R\Delta^2$ when we choose the positive sign but $\Delta < 0$. The choice of the sign is based on the sign of the balance function,

$$Q_- - Q_+ = \mathrm{E}\left\{L_-(\widehat{\Delta}; \Delta); \delta, q\right\} - \mathrm{E}\left\{L_+(\widehat{\Delta}; \Delta); \delta, q\right\}.$$

With $\widehat{\Delta}$ realised, this is a function of δ and q. A prior distribution is called equilibrium if its parameter pair (δ, q) is a root of the balance function. For such a prior, the choice of the sign is immaterial. Below we evaluate the expected losses Q_- and Q_+, and although we fail to find a closed-form expression for the equilibria (δ, q), we show that there is a unique equilibrium for each q. The equilibrium function $\delta_0(q)$, which assigns to every q the corresponding equilibrium value of δ, has the following property. For a prior located below the function $\delta_0(q)$ the negative sign and for one above $\delta_0(q)$ the positive sign is associated with smaller expected loss.

The posterior distribution of Δ is $\mathcal{N}(\widetilde{\Delta}, S^2)$, where

$$\tilde{\Delta} = \frac{m\delta + q\hat{\Delta}}{m + q}$$

$$S^2 = \frac{mq}{m + q}\sigma^2,$$

derived similarly to (4.2). Denote $\tilde{z} = \tilde{\Delta}/S$. If we choose the negative sign the posterior expection of the piecewise quadratic loss is

$$
\begin{aligned}
Q_- &= \int_0^{+\infty} \frac{\Delta^2}{S}\,\phi\left(\frac{\Delta - \tilde{\Delta}}{S}\right)\,\mathrm{d}\Delta \\
&= \int_{-\tilde{z}}^{+\infty} \left(\tilde{\Delta} + Sz\right)^2 \phi(z)\,\mathrm{d}z \\
&= \tilde{\Delta}^2\{1 - \Phi(-\tilde{z})\} + 2S\tilde{\Delta}\phi(\tilde{z}) - S^2\tilde{z}\,\phi(\tilde{z}) + S^2\{1 - \Phi(-\tilde{z})\} \\
&= \left(\tilde{\Delta}^2 + S^2\right)\Phi(\tilde{z}) + S^2\tilde{z}\phi(\tilde{z}),
\end{aligned}
$$

obtained by operations similar to those applied in Sect. 2.1. If we choose the positive sign, we have

$$Q_+ = R\left(\tilde{\Delta}^2 + S^2\right)\{1 - \Phi(\tilde{z})\} - RS^2\tilde{z}\phi(\tilde{z}), \tag{4.4}$$

so the balance function, $\Delta Q = Q_- - Q_+$, is

$$\Delta Q = S^2\left[(R+1)\left\{\left(1 + \tilde{z}^2\right)\Phi(\tilde{z}) + \tilde{z}\phi(\tilde{z})\right\} - R\left(1 + \tilde{z}^2\right)\right]. \tag{4.5}$$

We search for the root of the scaled balance function $f(\tilde{z}) = \Delta Q/S^2$, which depends only on $\tilde{z}$. The derivative of this function is

$$\frac{\partial f}{\partial \tilde{z}} = 2(R+1)\{\tilde{z}\Phi(\tilde{z}) + \phi(\tilde{z})\} - 2R\tilde{z}.$$

Further differentiation yields the identity

$$\frac{\partial^2 f}{\partial \tilde{z}^2} = 2(R+1)\Phi(\tilde{z}) - 2R.$$

This is an increasing function of $\tilde{z}$, with a sole root at $\tilde{z} = \Phi^{-1}\{R/(R+1)\}$. The first-order derivative attains its minimum at this value of $\tilde{z}$, and the minimum is equal to $2(R+1)\phi(\tilde{z})$. Since this is positive, $\partial f/\partial \tilde{z}$ is positive throughout, and so f is an increasing function of $\tilde{z}$. It is easy to check that f has limits $\pm\infty$ as $\tilde{z} \to \pm\infty$. Therefore, f has a unique minimum, and it is at the root of $\partial f/\partial \tilde{z}$. For fixed q, $\tilde{z}$ is a linear function of δ, and so there is a unique equilibrium prior (δ, q) for each $q > 0$. Thus, the equilibria can be described by a function $\delta_0(q)$.

In a practical problem, we proceed by the following steps. Having obtained the difference of the sample means, $\widehat{\Delta}$, we find the root of the balance function (4.5), denoted by z^*, and solve the equation

$$z^* = \frac{\widetilde{\Delta}}{S} = \frac{m\delta + q\widehat{\Delta}}{\sigma\sqrt{mq(m+q)}}$$

for δ as a function of q. The solution is

$$\delta_0(q) = \frac{z^*\sigma\sqrt{mq(m+q)} - q\widehat{\Delta}}{m}. \tag{4.6}$$

If we have already elicited a (single) prior from the expert, then we relate its parameters (δ, q) to the equilibrium function. If (δ, q) lies underneath $\delta_0(q)$, the negative sign is preferred; if it lies above, positive sign is preferred. If a set of plausible prior parameter vectors (δ, q) is declared, and the entire set lies on one side of the equilibrium function δ_0, below or above it, we have the same (common) conclusion for every one of them. See the left-hand panel of Fig. 4.1 for an illustration. If the function intersects the plausible set an unequivocal decision cannot be made, because for some priors one sign and for others the other sign would be preferred; see the right-hand panel. One resolution of the problem is to continue the elicitation and attempt to reduce the set of plausible priors. However, if the integrity of the process is breached and a plausible set smaller than what is warranted is declared, the unequivocality of the decision, although desired and convenient, may not be justified.

The piecewise linear loss function is defined as Δ when we choose the negative sign but $\Delta > 0$, and as $-R\Delta$ when we choose the positive sign but $\Delta < 0$. The balance function $\Delta Q(\tilde{z}) = Q_- - Q_+$ for this loss function is

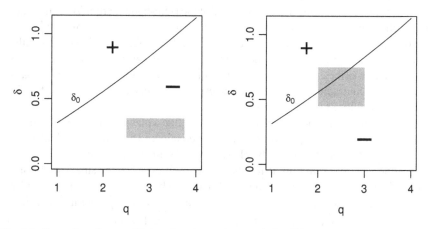

Fig. 4.1 Examples of an equilibrium function and a set of plausible priors. The set of priors is represented by the shaded rectangle in each panel. *Left-hand panel* unequivocal decision (*negative sign*); *right-hand panel* no clear (equivocal) decision

$$\Delta Q(\tilde{z}) = S\left[R\tilde{z} - (R-1)\{\tilde{z}\Phi(\tilde{z}) + \phi(\tilde{z})\}\right], \tag{4.7}$$

derived similarly to (4.5). Its derivative is $S\{(R-1)\Phi(\tilde{z}) + R\}$. For the piecewise constant loss, equal to one unit when we incorrectly choose the negative sign and to R when we incorrectly choose the positive sign, the balance function is $\Delta Q(\tilde{z}) = (R+1)\Phi(\tilde{z}) - R$, so its root is $z^* = \Phi^{-1}\{R/(R+1)\}$.

The parameter q is the prior version of $m = 1/n_1 + 1/n_2$. When the data contain much more information than the prior and $q \gg m$, $\sqrt{q(m+q)} \doteq q + \frac{1}{2}m$ and (4.6) can be approximated as

$$\delta_0(q) \doteq \frac{z^*\sigma\left(q + \frac{1}{2}m\right)}{\sqrt{m}} - \frac{q\widehat{\Delta}}{m};$$

$\delta_0(q)$ is close to a linear function (see Fig. 4.1) and its properties are easy to describe. For example, when extended to negative values of q, the equilibrium functions for different values of σ^2 converge to a point near $q_0 = -\frac{1}{2}m$.

Example

Suppose in a study to compare two medical treatments we obtained $\widehat{\Delta} = 0.75$ with samples of sizes $n_1 = 20$ and $n_2 = 25$, so that $m = 0.09$. Suppose the plausible range of values of q is $(0.25, 0.75)$. The assumption that σ^2 is known is rarely realistic. Here we compensate for it by evaluating the equilibrium functions for $\sigma^2 \in (0.6, 1.5)$. For piecewise quadratic and linear loss functions, we evaluate $\delta_0(q)$ by the Newton-Raphson algorithm. For the piecewise absolute loss, the solution has a closed form.

The equilibrium functions are drawn in Fig. 4.2 for the three classes of loss functions (rows) and penalty ratios $R = 25$ and 100 (columns). In each panel, the function $\delta_0(q)$ is drawn for the values of σ^2 indicated at the right-hand margin. As anticipated, the equilibrium functions are very close to linearity, and are increasing in σ^2.

For $R = 25$ with the piecewise quadratic and linear loss functions, the equilibrium functions attain negative values. Thus, the positive sign is preferred even with some negative prior expectations δ; the recorded data contains ample information to decisively contradict prior information, especially when q is large (close to 0.75) and σ^2 small (close to 0.6).

If we are uncertain about the value of σ^2 we define a plausible range for it, such as $(0.6, 1.5)$. We choose the positive sign only when it would be chosen for all plausible values of σ^2, that is, if it would be chosen even for their maximum, $\sigma^2 = 1.5$. We choose the negative sign only if it would be chosen even for the smallest plausible value of σ^2, equal to 0.6. In all other scenarios, when the positive sign would be chosen for some plausible values of σ^2, say, for $\sigma^2 \in (0.6, \sigma_{\dagger}^2)$, but not for others, $\sigma^2 \in (\sigma_{\dagger}^2, 1.5)$, we reach an impasse, no clear choice, unless we can narrow down the plausible range of σ^2. The region bounded by the equilibrium functions for $\sigma^2 = 0.6$ and $\sigma^2 = 1.5$ can be regarded as a *gray zone*, in which the choice of the sign is

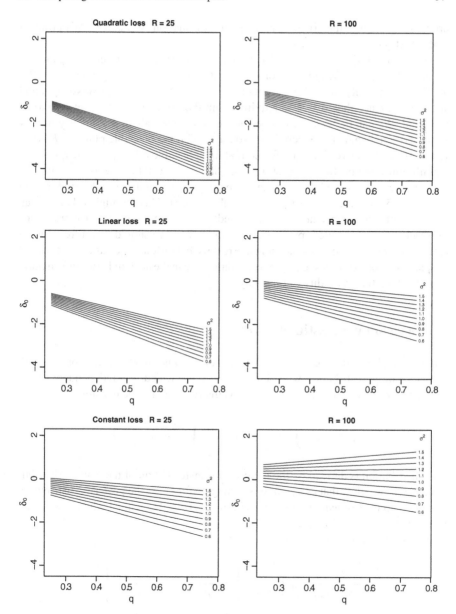

Fig. 4.2 Equilibrium functions for comparing two samples with normal distributions for quadratic, linear and absolute loss functions and penalty ratios $R = 25$ and 100; sample sizes $n_1 = 20$ and $n_2 = 25$ and estimate $\widehat{\Delta} = 0.75$

problematic. This is the price for incomplete information. We address this problem in the next section.

For $R = 100$ with all three loss functions and values of σ^2 (the right-hand panels of Figure 4.2), the equilibrium functions are greater than their counterparts for $R = 25$. With higher penalty ratio R we are more averse to choosing the positive sign, and stronger evidence that $\Delta > 0$ is required from the prior for the positive sign to be preferred. We can define a gray zone for the uncertainty about R similarly to how we propose to handle the uncertainty about σ^2. With the plausible range of R set to $(25, 100)$, the positive sign is preferred if all the plausible priors (δ, q) lie above the equilibrium function for $R = 100$ (and the largest plausible value of σ^2), and the negative sign is preferred if all the plausible priors lie under the equilibrium function for $R = 25$ (and the smallest plausible value of σ^2). This highlights that we can operate with uncertainty about both σ^2 and R, at the price of a wider gray zone in which impasse, no clear preference, is concluded. We should therefore strive to reduce the gray zone to avoid inconclusive results, but only as much as the integrity of the parties involved allows. Unsupported confidence in the declared prior information would deflate the credibility of the analysis.

4.3 Decision with Estimated σ^2

Suppose in the setting of the previous section we also have a prior for σ^2, with inverse gamma distribution with shape ν and scale s, and it is independent of the prior $\mathcal{N}(\delta, q\sigma^2)$ for Δ. The inverse gamma distribution has the density

$$f\left(\sigma^2 \mid \nu, s\right) = \frac{1}{\Gamma(\nu)}\, s^{\nu} \left(\frac{1}{\sigma^2}\right)^{\nu+1} \exp\left(-\frac{s}{\sigma^2}\right).$$

This distribution is derived by the reciprocal transformation of the gamma distribution. We show below that as a prior it is conjugate for σ^2 — the posterior distribution of σ^2 is also inverse gamma. Its expectation is $s/(\nu - 1)$, defined when $\nu > 1$, and variance $s^2/(\nu - 1)^2/(\nu - 2)$, when $\nu > 2$.

The joint posterior density of Δ and σ^2 is

$$\frac{C}{2\pi\sigma^2\sqrt{mq}} \exp\left[-\frac{1}{2\sigma^2}\left\{\frac{(\Delta - \widehat{\Delta})^2}{m} + \frac{(\Delta - \delta)^2}{q}\right\}\right]$$
$$\times \frac{s^{\nu}}{\Gamma_2(k)\,\Gamma(\nu)} \left(\frac{k}{2}\right)^{k/2} \left(\frac{1}{\sigma^2}\right)^{k/2+\nu+1} \hat{\sigma}^{k-2} \exp\left\{-\frac{1}{2\sigma^2}\left(k\hat{\sigma}^2 + 2s\right)\right\},$$

where $k = n_1 + n_2 - 1$; recall that $\Gamma_2(x) = \Gamma(\frac{1}{2}x)$. We consolidate all the arguments of the exponentials into $\exp\{-\frac{1}{2}G(\Delta)/\sigma^2\}$, where

$$G(\Delta) = u\left(\Delta - \widetilde{\Delta}\right)^2 + \frac{1}{m+q}\left(\delta - \widehat{\Delta}\right)^2 + k\widehat{\sigma}^2 + 2s \tag{4.8}$$

and $u = (m+q)/(mq)$. The remainder of the expression is the factor $1/\sigma^{2(k/2+\nu+2)}$ and another factor, denoted by C', which depends on neither Δ nor σ^2. The first term in (4.8) does not involve $\widehat{\sigma}^2$, ν or s; it can be interpreted as an estimator of σ^2 informed solely by the prior for Δ. It is associated with a single degree of freedom. From the remainder of $G(\Delta)$, we can define the estimator

$$\widetilde{\sigma}^2 = \frac{1}{h}\left\{\frac{1}{m+q}\left(\delta - \widehat{\Delta}\right)^2 + k\widehat{\sigma}^2 + 2s\right\},$$

where $h = k + 2\nu + 1 = n_1 + n_2 + 2\nu - 1$, which involves the prior information about σ^2. The data-related k degrees of freedom in $\widehat{\sigma}^2$ are supplemented by 2ν from the prior $(2s)$ and one from $(\delta - \widehat{\Delta})^2$. The ratio $\sigma_\pi^2 = s/\nu$ can be interpreted as a prior estimate of the variance σ^2.

The joint posterior density of Δ and σ^2 is

$$g\left(\Delta, \sigma^2\right) = C'\left(\frac{1}{\sigma^2}\right)^{k/2+\nu+2}\exp\left\{-\frac{G(\Delta)}{2\sigma^2}\right\}.$$

Inferences related exclusively to Δ or σ^2 are based on the corresponding marginal posterior distributions. By extracting from $g(\Delta, \sigma^2)$ the density of $\mathcal{N}(\widetilde{\Delta}, \sigma^2/u)$, we obtain the marginal posterior distribution of σ^2:

$$\int_{-\infty}^{+\infty} g\left(\Delta, \sigma^2\right) d\Delta = D\left(\frac{1}{\sigma^2}\right)^{h/2+1}\exp\left\{-\frac{h\widetilde{\sigma}^2}{2\sigma^2}\right\},$$

where D is the constant for which the right-hand side is a density. Thus, the posterior distribution of σ^2 is inverse gamma, with shape $\frac{1}{2}h$ and scale $\frac{1}{2}h\widetilde{\sigma}^2$.

The marginal posterior of Δ is derived by integrating the joint posterior density over σ^2;

$$\int_0^{+\infty} g\left(\Delta, \sigma^2\right) d\sigma^2 = C''\{G(\Delta)\}^{-(k/2+\nu+1)}, \tag{4.9}$$

obtained by matching the density $g(\Delta, \sigma^2)$, as a function of σ^2, with Δ fixed, to the density of an inverse gamma distribution. Denote by ψ_h the density of the t distribution with h degrees of freedom,

$$\psi_h(x) = \frac{\Gamma_2(h+1)}{\Gamma_2(h)}\frac{1}{\sqrt{h\pi}}\left(\frac{1}{1+x^2/h}\right)^{h/2+1/2}, \tag{4.10}$$

and by Ψ_h its distribution function. The distribution is symmetric, $\psi_h(x) = \psi_h(-x)$ for all x, so its expectation is zero when it is defined (for $h > 1$). The variance of the

t distribution is $h/(h-2)$ for $h > 2$; for $h \leq 2$ it is not defined. Let $\gamma_h = \sqrt{1-2/h}$; this is the reciprocal of the standard deviation.

The function in (4.9) is the density of a scaled non-central t distribution with $h = k + 2\nu + 1$ degrees of freedom. That is, the variable $\xi = \sqrt{u}(\Delta - \tilde{\Delta})/\tilde{\sigma}$ has the central t distribution with h degrees of freedom. To evaluate the balance function for Δ with the piecewise quadratic loss function, we apply the identity

$$\frac{\partial}{\partial t} \frac{\psi_{h-2}(t\gamma_h)}{\gamma_h} = -t\psi_h(t). \tag{4.11}$$

It is derived directly from the definition of ψ in (4.10). As $h \to +\infty$, ψ_h converges pointwise to the density of the standard normal distribution. By taking the limit $h \to +\infty$ in (4.11), we also obtain the corresponding identity for the standard normal, $\partial\phi/\partial x = -x\phi(x)$, used in Chap. 2. Let $S_t = \tilde{\sigma}/\sqrt{u}$ and $\tilde{t} = \tilde{\Delta}/S_t$. These terms are the counterparts of S and $\tilde{z}$ used in Sect. 4.2. Note that δ is involved in S_t, whereas in S it is not.

The balance function for Δ with the quadratic loss function is

$$
\begin{aligned}
\Lambda(\tilde{t}) &= \int_0^{+\infty} \Delta^2 \psi_h\left(\frac{\Delta - \tilde{\Delta}}{S_t}\right) d\Delta - R\int_{-\infty}^0 \Delta^2 \psi_h\left(\frac{\Delta - \tilde{\Delta}}{S_t}\right) d\Delta \\
&= \int_{-\tilde{t}}^{+\infty} \left(\tilde{\Delta} + S_t t\right)^2 \psi_h(t)\,dt - R\int_{-\infty}^{-\tilde{t}} \left(\tilde{\Delta} + S_t t\right)^2 \psi_h(t)\,dt \\
&= \tilde{\Delta}^2 \left\{(R+1)\Psi_h(-\tilde{t}) - R\right\} \\
&\quad - \frac{2S_t\tilde{\Delta}}{\gamma_h}\left\{\left[\psi_{h-2}(t\gamma_h)\right]_{-\tilde{t}}^{+\infty} + R\left[\psi_{h-2}(t\gamma_h)\right]_{-\infty}^{-\tilde{t}}\right\} \\
&\quad - \frac{S_t^2}{\gamma_h}\left\{\left[t\psi_{h-2}(t\gamma_h)\right]_{-\tilde{t}}^{+\infty} + R\left[t\psi_{h-2}(t\gamma_h)\right]_{-\infty}^{-\tilde{t}}\right\} \\
&\quad + \frac{S_t^2}{\gamma_h}\left\{\int_{-\tilde{t}}^{+\infty} \psi_{h-2}(t\gamma_h)\,dt - R\int_{-\infty}^{-\tilde{t}} \psi_{h-2}(t\gamma_h)\,dt\right\} \\
&= S_t^2\left[(R+1)\left\{\tilde{t}^2\Psi_h(\tilde{t}) + \frac{1}{\gamma_h^2}\Psi_{h-2}(\tilde{t}\gamma_h) + \frac{\tilde{t}}{\gamma_h}\psi_{h-2}(\tilde{t}\gamma_h)\right\} - R\left(\tilde{t}^2 + \frac{1}{\gamma_h^2}\right)\right],
\end{aligned}
$$

obtained by the same steps as its normality counterpart in (4.5). We find the root of Λ by the Newton-Raphson algorithm, for which we use the following expression:

$$\frac{\partial\Lambda}{\partial\tilde{t}} = 2S_t^2\left[(R+1)\left\{\tilde{t}\,\Psi_h\left(\tilde{t}\right) + \frac{\psi_{h-2}\left(\tilde{t}\gamma_h\right)}{\gamma_h}\right\} - R\tilde{t}\right].$$

The second-order derivative is

$$\frac{\partial^2 \Lambda}{\partial \tilde{t}^2} = 2S_t^2 \left\{ (R+1)\Psi_h(\tilde{t}) - R \right\}.$$

Proving that Λ has a single (unique) root, by adapting the corresponding proof for normally distributed outcomes, is left for an exercise.

The balance function for the piecewise linear loss is

$$\Lambda(\tilde{t}) = S_t \left[R\tilde{t} - (R-1) \left\{ \tilde{t}\Psi_h(\tilde{t}) + \frac{1}{\gamma_h} \psi_{h-2}(\tilde{t}\gamma_h) \right\} \right]$$

with

$$\frac{\partial \Lambda}{\partial \tilde{t}} = S_t \left\{ R - (R-1)\Psi_h(\tilde{t}) \right\}.$$

For piecewise constant loss the root of the balance function,

$$\Lambda(\tilde{t}) = (R+1)\Psi_h(\tilde{t}) - R,$$

is $t^* = \Psi_h^{-1}\{R/(R+1)\}$.

The equilibrium function is constructed by first finding the root t^* of the balance function and then solving the equation

$$t^* = \frac{\tilde{\Delta}}{S_t}$$

for the prior parameters δ, q, s and v. The latter task, with four unknowns, is too complex. We simplify it by solving it for fixed $\sigma_\pi^2 = s/v_2$ and $v_2 = 2v$, in the plausible ranges of their values. We prefer to use these two parameters because elicitation of their values is easier with reference to their interpretation as a prior estimate of σ^2 and the associated number of degrees of freedom.

The equilibrium functions for the setting of the example in the previous section (see Fig. 4.2) are displayed in Fig. 4.3. Additionally, we assume that $\hat{\sigma}^2 = 1.22$, $v \in (8, 12)$, $\sigma_\pi^2 = (0.875, 1.3)$ and $q \in (0.25, 0.75)$. The values of the equilibrium function are monotone in v and σ_π^2, so it suffices to evaluate the function for the end points of the intervals of their plausible values. The four lines in each panel correspond to these configurations of v and σ_π^2. They are not exactly linear, but their curvature is only slight. The space delimited by the lines, a thin wedge in each panel, is the gray zone, in which the decision is equivocal, where an impasse is reached.

The expressions for the balance of the losses resemble their counterparts in Sect. 4.2, but do not allow as simple a discussion as when σ^2 is known, because S_t is more complex than S. Also, the balance equation for the piecewise quadratic loss involves two distribution functions, t with h and $h-2$ degrees of freedom, instead of a single normal distribution.

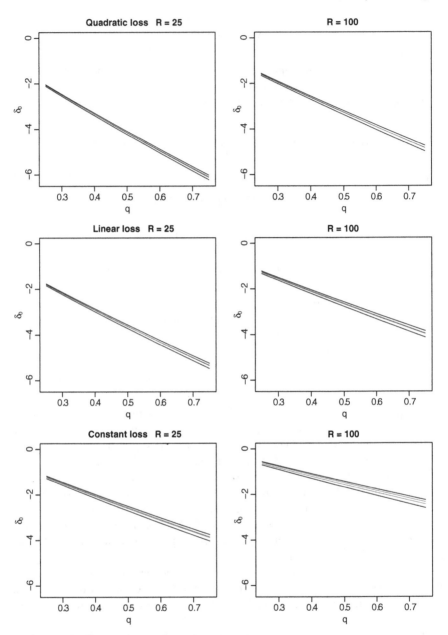

Fig. 4.3 Equilibrium functions for quadratic, linear and absolute loss functions and penalty ratios $R = 25$ and 100, with estimated variance σ^2

4.4 Problems, Exercises and Suggested Reading

1. Find the conjugate prior distributions for the parameters of the exponential, Poisson, binomial and beta distributions. Apply the Bayes theorem in these settings.
2. Interpret the prior distributions in the previous example as representing prior observations.
3. Find a set of sufficient statistics for the parameters of a beta distribution. Do the same for the gamma distribution.
4. Show on examples in the text that as the number of observations increases and the prior distribution is held fixed, the influence of the prior on the posterior diminishes.
5. Discuss how the methods in Sect. 4.2 can be adapted to comparing the expectation of a normal random sample with a constant.
6. Discuss the advantage of estimating the normal-sample variance σ^2 for inferences about the mean μ, as compared to a sensitivity analysis with a range of plausible values of σ^2. How would you assess the value of prior information for deciding about the sign of Δ in the setting of Sect. 4.3?
7. A project: Work out the details for estimating the ratio of two variances from independent normally distributed random samples. Define suitable (conjugate) priors for the variances. Formulate the problem of deciding whether the ratio is greater or smaller than 1.0 in such a way that its solution would be tractable. Assistance: Longford and Andrade Bejarano (2013).
8. A project: Discuss how you would adapt the methods in this chapter to several samples. That is, suppose we have K samples, and would like to choose the one with the highest (or lowest) expectation.
9. A project: Work out the details for comparing the expectations of the elements of a bivariate normal distribution. Assume first that the (2×2) variance matrix is known, and then search for weaker (and more realistic) assumptions for which the problem is still manageable, in the frequentist or Bayesian paradigm.
10. A project: Work out the details for comparing the expectations of two scaled t distributions (with known scales). Formulate a guideline for when the results with t distributions are indistinguishable from the results of the analysis with the assumptions of normality.
11. For more details about the analysis in Sect. 4.3, see Longford (2012).
12. Suggested reading on Bayesian analysis: Berry (1996), Gelman et al. (2003), and Lee (2004), Carlin and Louis (2008) and Albert (2009). About Bayesian decision theory: Smith (1988) and Robert (2007).
13. Suggested reading about elicitation: O'Hagan (1998), Garthwaite and O'Hagan (2000), Garthwaite et al. (2005).

References

Albert, J. (2009). *Bayesian computation with R* (2nd ed.). New York: Springer.

Berry, D. (1996). *Statistics: A Bayesian perspective*. Belmont: Duxbury.

Carlin, B. P., & Louis, T. A. (2008). *Bayesian methods for data analysis* (3rd ed.). Boca Raton: Chapman and Hall/CRC.

Garthwaite, P. H., & O'Hagan, A. (2000). Quantifying expert opinion in the UK water industry: An experimental study. *Journal of the Royal Statistical Society Ser D, 49,* 455–477.

Garthwaite, P. H., Kadane, J. B., & O'Hagan, A. (2005). Statistical methods for eliciting probability distributions. *Journal of the American Statistical Association, 100,* 680–701.

Gelman, A., Carlin, J. B., Stern, H. S., & Rubin, D. B. (2003). *Bayesian data analysis* (2nd ed.). Boca Raton: Chapman and Hall/CRC.

Lee, P. M. (2004). *Bayesian statistics: An introduction* (3rd ed.). Chichester: Wiley.

Longford, N. T. (2012). Comparing normal random samples, with uncertainty about the priors and utilities. *Scandinavian Journal of Statistics, 39,* 729–742.

Longford, N. T., & Andrade Bejarano, M. (2013). Decision theory for the variance ratio in ANOVA with random effects. Submitted.

O'Hagan, A. (1998). Eliciting expert beliefs in substantial practical applications. *Journal of the Royal Statistical Society Ser D, 47,* 21–68.

Robert, C. (2007). *The Bayesian choice: From decision-theoretic foundations to computational implementation.* New York: Springer.

Smith, J. Q. (1988). *Decision analysis. A Bayesian approach.* New York: Oxford University Press.

Chapter 5
Data from Other Distributions

In this chapter we look beyond the normal (and t) distributions and adapt the methods of the previous chapters for some other common discrete and continuous distributions. Although the development is presented in the Bayesian perspective, it can be adapted for the frequentist by replacing the prior distribution with prior data.

5.1 Binary Outcomes

For a random sample of large size from a binary distribution with a probability distant from zero and unity, we can refer to the normal approximation and apply the methods presented in Chaps. 2 and 4. With small samples and extreme probabilities, the conventional approaches are easily shown wanting. For example, the sampling variance of the sample proportion $\hat{p}$ is usually estimated naively, by $\hat{p}(1 - \hat{p})/n$. When $\hat{p} = 0$, it is outright misleading, because it indicates absence of any uncertainty. The source of the problem is the nonlinear transformation of p to $p(1 - p)$; $\hat{p}^2$ may be a poor estimator of p^2 even when $\hat{p}$ is efficient for p. For very small or very large p, $p(1 - p)$ is close to a linear function of p, namely, to p or $1 - p$, so the error committed in estimating p reappears when estimating $\mathrm{var}(\hat{p})$.

In this section, we focus on the problem of choosing between two courses of action, A and B; A is appropriate when p is small, say, smaller than p_0, and B when $p > p_0$. For example, a manufacturer or another professional would derive some kudos (custom, good reputation or profit) from the claim that an item it produces is faulty with a probability smaller than the established standard of p_0. At first sight, this might be an obvious application of hypothesis testing, especially if we resolve the problems with estimating $\mathrm{var}(\hat{p})$, highlighted earlier. We would like to tailor the solution for a particular client, such as a manufacturer's representative who is familiar with the production process and the market for its product, and can describe and maybe quantify the consequences of making a claim that would later turn out to be false, and of not making the claim even though making it would be appropriate.

N. T. Longford, *Statistical Decision Theory*,
SpringerBriefs in Statistics, DOI: 10.1007/978-3-642-40433-7_5,
© The Author(s) 2013

In the outlined setting, a false claim would be punished more harshly in terms of reduced sales, lost reputation, and the like, than excessive modesty (disadvantage in new markets in particular). Clients may have different perspectives and priorities, one keener to gamble with the reputation and another more cautious with any public statement, and in advertising—one solution does not fit everybody.

A binomial trial with n attempts is defined as a sequence of n independent attempts, each with a binary outcome (no and yes, 0 and 1, or failure and success) with the same probability of failure p. Suppose such a trial yielded k failures in n attempts. Is the underlying probability of a failure smaller than p_0? We add to this statement of the problem two items of prior (background) information. First, refraining from the claim when it could have been made (the false negative statement) incurs unit loss, whereas the false affirmative statement is associated with loss $R > 1$. Next, a beta distribution is declared as the prior for p. The class of beta distributions is given by the densities

$$f(x; \alpha, \beta) = \frac{\Gamma(\alpha + \beta)}{\Gamma(\alpha)\,\Gamma(\beta)}\, x^{\alpha-1}(1-x)^{\beta-1}$$

for $x \in (0, 1)$, with parameters $\alpha > 0$ and $\beta > 0$. The distributions have expectations $p_\pi = \alpha/(\alpha + \beta)$ and variances $p_\pi(1 - p_\pi)/(\alpha + \beta + 1)$.

The posterior density of the probability p is equal to the product of the data probability and the prior density, standardised to be a density:

$$C p^k (1-p)^{n-k} p^{\alpha-1}(1-p)^{\beta-1} = \frac{\Gamma(n+\alpha+\beta)}{\Gamma(k+\alpha)\,\Gamma(n-k+\beta)}\, p^{k+\alpha-1}(1-p)^{n-k+\beta-1},$$

where C is the standardising constant, obtained by matching the rest of the expression to a beta distribution. The prior parameters α and β can be interpreted as the respective prior numbers of failures and successes in $\alpha + \beta$ attempts, additional to the k failures and $n - k$ successes observed in n attempts. This description or interpretation of the prior is invaluable for elicitation of the parameters α and β, because a client may be more comfortable with relating his prior information to outcomes of hypothetical trials (of the same nature as the realised trial) than to density functions.

Denote by $B(y; \alpha, \beta)$ the distribution function of the beta distribution with parameters α and β. The (posterior) expected loss with the piecewise constant loss function has the components

$$\begin{aligned}
L_A &= \frac{R\,\Gamma(n+\alpha+\beta)}{\Gamma(k+\alpha)\,\Gamma(n-k+\beta)} \int_{p_0}^{1} p^{k+\alpha-1}(1-p)^{n-k+\beta-1}\,\mathrm{d}p \\
&= R\,\{1 - B(p_0; k+\alpha, n-k+\beta)\} \\
L_B &= \frac{\Gamma(n+\alpha+\beta)}{\Gamma(k+\alpha)\,\Gamma(n-k+\beta)} \int_{0}^{p_0} p^{k+\alpha-1}(1-p)^{n-k+\beta-1}\,\mathrm{d}p \\
&= B(p_0; k+\alpha, n-k+\beta),
\end{aligned}$$

when we choose action A (make the claim) or action B (refrain from the claim), respectively. For a given beta prior, specified by α and β, penalty ratio R and outcome k (out of n attempts), we choose the action with the smaller expected loss. If the study is to take place, with the number of attempts n set, we can declare in advance the rule for selecting the action based on the number of successes k. The equilibrium, when $L_A = L_B$ and the choice of the action is immaterial, corresponds to the equation

$$B\left(p_0 ; k + \alpha, n - k + \beta\right) = \frac{R}{R+1} . \qquad (5.1)$$

We can solve this equation in a number of ways. For a given outcome, prior and penalty ratio R, we can find the equilibrium threshold p_0^* for which we would be indifferent as to whether to claim that $p < p_0^*$ or not. By solving it for R, with the other factors fixed, we establish how harsh a penalty for the claim should make us indifferent to the choice. For fixed outcome k, p_0 and R, we can find the equilibrium priors, for which the choice is immaterial.

Suppose an experiment with a binomial outcome concluded with two failures in 125 attempts. We would like to establish whether the rate of faults is lower than 2%. We set the penalty ratio to 10, but will study the results for $R \in (5, 20)$. The balance equation in (5.1) is solved numerically. We solve it on a grid of plausible values of α, finding the corresponding equilibrium $\beta_R(\alpha)$. For $\alpha = 0$, we find, in effect, how many additional successes we need for the claim that $p < p_0$ to make sense. Of course, the solution may be negative, indicating that we could have made the claim even if we had fewer than 123 successes with the two failures. Therefore, the equilibrium function may be well defined only for $\alpha > \alpha_0$, or $\beta_R(\alpha)$ is positive even for $\alpha = 0$.

It is more practical to solve the balance equation (5.1) by the Newton method, because we do not have a convenient expression for the derivative of the beta density with respect to its parameters α and β. The solution is presented in the left-hand panel of Fig. 5.1 in the form of the equilibrium functions $\beta_R(\alpha)$ for $R = 5$, 10 and 20. The right-hand panel contains the plot of the equilibrium prior probability $p_\pi = \alpha / \{\alpha + \beta_R(\alpha)\}$ against the equilibrium prior sample size $n_\pi = \alpha + \beta_R(\alpha)$.

For a selected value of R, we make the claim if the prior lies below the equilibrium function in the (n, p) parametrisation (above in the (α, β) parametrisation), and refrain from making the claim otherwise. It is easier to work with $\beta_R(\alpha)$, which has very little curvature, even though the equilibrium functions in the (n, p) parametrisation, $n_R(p)$ and $p_R(n)$, may appeal to a client better. For example, the elicitation may start with the discussion of the prior sample size as the worth of the information expressed in terms of the number of attempts, and then proceed to the number of successes (possibly fractional) in these 'prior' attempts.

We can interpret the result in Fig. 5.1 as follows. In order to make a justified claim, we require an 'optimistic' prior that supports the claim. About $\beta_5(0) \doteq 36$ successes and no failures have to be added to the realised attempts (2 out of 125) for the claim to make sense with $R = 5$, and $\beta_{20}(0) = 114$ successes have to be added with $R = 20$.

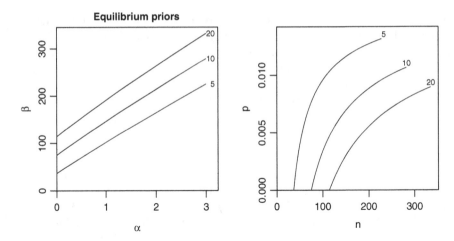

Fig. 5.1 The equilibrium priors for a binomial trial with the result of two failures in 125 attempts. The threshold is $p_0 = 0.02$; the penalty ratios $R = 5$, 10 and 20 are marked at the right-hand end of each curve

The slope of the functions β_R is about 60. Thus, every additional prior failure would have to be accompanied with about 60 successes.

If we are uncertain about R, as when the plausible range of the penalty ratios is $(R_L, R_U) = (5, 20)$, then the strip delimited by the equilibrium functions $\beta_5(\alpha)$ and $\beta_{20}(\alpha)$ is a gray zone. With any prior (α, β) in this zone, we reach an impasse because the decision to claim or not is equivocal; for some plausible values of R one action and for the complement of values the other action is preferred. When $R_L = R_U$, the gray zone degenerates to the equilibrium function (curve) but, when convenient, we can still refer to it as a (very narrow) gray zone.

Uncertainty about the prior is treated similarly. We represent the plausible priors by a set in the space (α, β) or (n, p). If this set has an overlap with the gray zone (or is intersected by the equilibrium function), then we reach an impasse, because for some plausible priors one action and for the complement the other action is preferred. Thus, it pays to have less uncertainty in the input to the analysis—the elicitation is a worthwhile activity, but its integrity is imperative. That is, if the client is not willing to rule out any prior or penalty ratio outside the plausible set, or has second thoughts about it, then the clearcut conclusion of the analysis may be poorly founded.

For piecewise linear loss we have the expected losses

$$L_A = R \int_{p_0}^{1} (p - p_0) f(p; k + \alpha, n - k + \beta) \, dp$$

$$L_B = \int_{0}^{p_0} (p_0 - p) f(p; k + \alpha, n - k + \beta) \, dp \, .$$

We make use of efficient algorithms for evaluating the distribution function of the beta by using the identity

$$p f(p; \alpha, \beta) = \frac{\alpha}{\alpha + \beta} f(p; \alpha + 1, \beta) \tag{5.2}$$

for any beta density f and $p \in (0, 1)$. With it, we obtain the balance equation

$$\frac{k + \alpha}{n + \alpha + \beta} B(p_0; k + \alpha + 1, n - k + \beta) - p_0 B(p_0; k + \alpha, n - k + \beta)$$
$$= \frac{R}{R - 1} \left(\frac{k + \alpha}{n + \alpha + \beta} - p_0 \right). \tag{5.3}$$

The equation is solved by the Newton method.

The expected piecewise quadratic loss involves integration of a function of the form $p^2 f(p)$. By reusing the identity in (5.2), we obtain

$$p^2 f(p; \alpha, \beta) = c_0 c_1 f(p; \alpha + 2, \beta), \tag{5.4}$$

where $c_h = (\alpha + h)/(\alpha + \beta + h)$, $h = 0, 1$. Therefore, the expected loss when we refrain from the claim is

$$L_B = \int_0^{p_0} (p - p_0)^2 f(p; k + \alpha, n - k + \beta) \, dp$$
$$= c_0 c_1 B(p_0; k + \alpha + 2, n - k + \beta) - 2 c_0 p_0 B(p_0; k + \alpha + 1, n - k + \beta)$$
$$+ p_0^2 B(p_0; k + \alpha, n - k + \beta).$$

The expected loss when we make the claim is

$$L_A = R \int_{p_0}^1 (p - p_0)^2 f(p; k + \alpha, n - k + \beta) \, dp$$
$$= R \Big\{ c_0 c_1 - 2 c_0 p_0 + p_0^2 - c_0 c_1 B(p_0; k + \alpha + 2, n - k + \beta)$$
$$+ 2 c_0 p_0 B(p_0; k + \alpha + 1, n - k + \beta) - p_0^2 B(p_0; k + \alpha, n - k + \beta) \Big\}.$$

Hence the balance equation

$$c_0 c_1 B(p_0; k + \alpha + 2, n - k + \beta) - 2 c_0 p_0 B(p_0; k + \alpha + 1, n - k + \beta)$$
$$+ p_0^2 B(p_0; k + \alpha, n - k + \beta) = \frac{R}{R + 1} \left(c_0 c_1 - 2 c_0 p_0 + p_0^2 \right),$$

which can be solved by the Newton method.

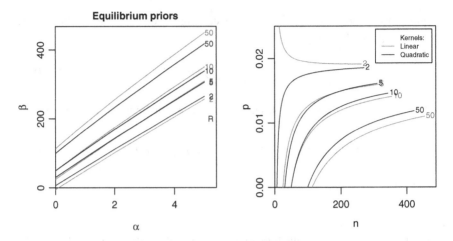

Fig. 5.2 The equilibrium priors for a binomial trial with the result of two failures in 125 attempts; loss functions with linear and quadratic kernel. The threshold is $p_0 = 0.02$ and the penalty ratios $R = 2, 5, 10$ and 50 are marked at the right-hand end of each curve

The equilibrium functions $\beta_R^{(L)}(\alpha)$ and $\beta_R^{(Q)}(\alpha)$ for the piecewise linear and quadratic loss (kernel) are plotted in Fig. 5.2 for a selection of penalty ratios R. Although the two sets of functions are drawn in the same diagram, we note that the kernels correspond to essentially different loss structures and their expected losses are not comparable, not even with the same penalty ratio R.

The functions $\beta_R^{(L)}$ and $\beta_R^{(Q)}$ have very little curvature and are all increasing. The functions $p_R^{(L)}(n) = \alpha/(\alpha + \beta_R^{(L)})$ and $p_R^{(Q)}(n) = \alpha/(\alpha + \beta_R^{(Q)})$ have steep slopes for small n, but for large n seem to settle down to asymptotes. In fact, this asymptote is $p = 0.02$, because for large n we have almost no uncertainty, the prior distribution is almost irrelevant, and the decision is straightforward. For $R = 2$, $p_2^{(L)}(n)$ decreases within the plotting range, unlike $p_2^{(Q)}(n)$, even though the corresponding functions $\beta_2^{(L)}(n)$ and $\beta_2^{(Q)}(n)$ differ only slightly. However, the latter functions attain small values for small α, so the ratio $\alpha/(\alpha + \beta)$ is rather unstable for them.

5.2 Poisson Counts

The Poisson distribution is commonly adopted as a model for count data. The distribution can be derived as the limit of the binomial when the number of attempts n diverges to $+\infty$ and the probability of failure (or success) p converges to zero in such a way that np has a finite positive limit, say, λ. The Poisson distribution is given by the probabilities

$$P(X = k) = \frac{e^{-\lambda}\lambda^k}{k!}$$

for $k = 0, 1, \dots$. The parameter $\lambda > 0$ is the expectation of X; the variance of X is also equal to λ. We assume that a study yields a random sample $k_1, \dots, k_n$ from a Poisson distribution with unknown value of λ. A sufficient statistic for λ is the sample mean $\bar{x} = (k_1 + \cdots + k_n)/n$.

The conjugate prior distribution for the Poisson is the gamma. We use the parametrisation for gamma distribution in terms of the rate ν and shape s. The density of the gamma distribution is

$$g(\lambda; \nu, s) = \frac{1}{\Gamma(s)} \nu^s \lambda^{s-1} \exp(-\lambda\nu).$$

Its expectation is s/ν and variance s/ν^2. Denote by G the distribution function of gamma. The posterior distribution of λ with this prior is also gamma, with the density

$$p(\lambda \mid \bar{x}; \nu, s) = C\lambda^{n\bar{x}+s-1} \exp\{-\lambda(n+\nu)\},$$

after gathering all the factors that do not involve λ in the constant term C. By matching the right-hand side with the appropriate gamma distribution, we establish that

$$C = \frac{(n+\nu)^{n\bar{x}+s}}{\Gamma(n\bar{x}+s)},$$

so the posterior distribution of λ is gamma with rate $n+\nu$ and shape $n\bar{x}+s$. We can interpret ν as the number of prior observations and s as the total of their values. As an alternative, we can use $\bar{s} = s/\nu$, as the prior version of $\bar{x}$.

Suppose we want to make a claim that $\lambda < \lambda_0$. As in the previous section, we assume that an unjustified claim (made when $\lambda > \lambda_0$) is associated with greater loss than the failure to make a claim when it would be justified, when $\lambda < \lambda_0$. Suppose first that the respective losses are $R > 1$ and unity, using a piecewise constant loss function. Then the equilibrium equation is

$$\int_0^{\lambda_0} g(\lambda; \nu+n, s+n\bar{x})\,d\lambda = R \int_{\lambda_0}^{+\infty} g(\lambda; \nu+n, s+n\bar{x})\,d\lambda,$$

that is,

$$G(\lambda_0; \nu+n, s+n\bar{x}) = \frac{R}{R+1}.$$

Compare this expression with (5.1) in the previous section.

For the linear and quadratic kernel loss, we require identities that avoid the direct integration of functions $xg(x; \nu, s)$ and $x^2g(x; \nu, s)$, as we have done by (5.2) and (5.4). By absorbing x or x^2 in the expression for the density, we obtain

$$xg(x; \nu, s) = \frac{s}{\nu} g(x; \nu, s + 1)$$

$$x^2 g(x; \nu, s) = \frac{s(s + 1)}{\nu^2} g(x; \nu, s + 2). \tag{5.5}$$

The derivation of these identities is left for an exercise. Note that integration of both sides of these identities leads to the expressions for the expectation and the expectation of the square of the distribution, and hence the variance. The balance equation for the linear kernel loss is

$$\int_0^{\lambda_0} (\lambda_0 - \lambda) g(\lambda; \nu + n, s + n\bar{x}) \, d\lambda = R \int_{\lambda_0}^{+\infty} (\lambda - \lambda_0) g(\lambda; \nu + n, s + n\bar{x}) \, d\lambda,$$

which, by exploiting the first identity in (5.5), reduces to

$$\frac{s + n\bar{x}}{\nu + n} G(\lambda_0; \nu + n, s + n\bar{x} + 1) - \lambda_0 G(\lambda_0; \nu + n, s + n\bar{x})$$

$$= \frac{R}{R - 1} \left(\frac{s + n\bar{x}}{\nu + n} - \lambda_0 \right).$$

Solving this equation is a task similar to its binomial counterpart in (5.3).

The balance equation for the quadratic kernel loss,

$$\int_0^{\lambda_0} (\lambda_0 - \lambda)^2 g(\lambda; \nu + n, s + n\bar{x}) \, d\lambda = R \int_{\lambda_0}^{+\infty} (\lambda - \lambda_0)^2 g(\lambda; \nu + n, s + n\bar{x}) \, d\lambda,$$

is equivalent to

$$\lambda_0^2 G(\lambda_0; \nu + n, s + n\bar{x}) - 2\lambda_0 d_0 G(\lambda_0; \nu + n, s + n\bar{x} + 1)$$

$$+ d_0 d_1 G(\lambda_0; \nu + n, s + n\bar{x} + 2) = \frac{R}{R + 1} \left(\lambda_0^2 - 2d_0\lambda_0 + d_0 d_1 \right),$$

where $d_h = (s + n\bar{x} + h)/(\nu + n)$, $h = 0, 1$. The equation is solved by the Newton method. The implementation for the binomial outcomes can be adapted for the Poisson with only minor changes.

Example

Table 5.1 displays the numbers of failures in 20 one-hour spells of uninterrupted (large-scale) production of an elementary item in the test of a manufacturing process after its fine-tuning. There were 4–14 failures, 7, 8 and 9 of them on three occasions each, and 172 failures in total.

The engineers want to decide whether the rate of faults is now below ten per hour, which is the established industrial standard. The fine-tuning is costly because of lost production, but failure to satisfy the standard would hurt the company in the long

Table 5.1 The numbers of failures in one-hour spells of an uninterrupted manufacturing process

Failures	4	5	6	7	8	9	10	11	12	14
Instances	1	1	2	3	3	3	2	3	1	1

term. Should the manufacturing process be fine-tuned further, or is its quality now satisfactory?

The engineers assessed the prior by the gamma distribution with rate $\nu = 7.5$ and shape $s = 75$ but, after a review prompted by concerns that this prior plays an important role in the analysis, agreed on the plausible range $6 \leq \nu \leq 10$ and $6 \leq s/\nu \leq 11$. Elicitation from the management concluded with the choice of the linear kernel loss with the plausible range of penalty ratios 20–50; the penalty for production of poor quality is certainly quite harsh, but the uncertainty about it is also considerable.

The data is summarised by its sample size $n = 20$ and sample mean $\bar{x} = 8.6$. The analysis is presented in Fig. 5.3. In the left-hand panel, the equilibrium functions for the three kernels are plotted on the scale ν versus s/ν, together with the plausible prior rectangle for $(\nu, s/\nu)$. The equilibrium functions are drawn for $R_L = 20$, $R = 35$ and $R_U = 50$.

We focus on the linear kernel loss (black curves); the curves for the absolute and quadratic kernel are drawn only for completeness. We can reduce our attention to the equilibrium function $s_{50}^{(L)}(\nu)$, because most of the rectangle lies under the gray zone formed by the plausible equilibrium functions, and a desirable outcome of the analysis would be if the entire rectangle was underneath all of them, that is,

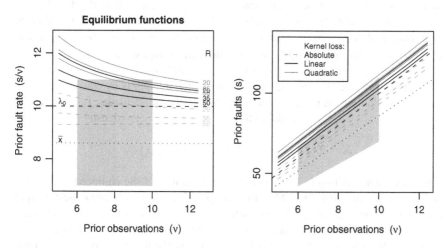

Fig. 5.3 The equilibrium priors for a sample of $n = 20$ Poisson counts with average outcome $\bar{x} = 8.6$ against the standard $\lambda_0 = 10.0$. Absolute, linear and quadratic kernel loss with penalty ratio $R = 20$, 35 and 50. The shaded areas in the two panels represent the plausible priors

under $s_{50}^{(L)}(\nu)$. The impasse due to this intersection could be resolved by reducing the plausible rectangle. In the particular exercise, this was regarded as out of the question because it would undermine the integrity of the management. Another solution is to continue with the trials. As n is increased, but if $\bar{x}$ remains unchanged at $8.6 < \lambda_0$, the equilibrium functions would increase because the stronger information about $\lambda < \lambda_0$ can be countered only by more pessimistic priors. However, it is by no means obvious how $\bar{x}$ would be changed by further trials. In the actual study, further ten observations, were made, with a total of 46 faults, and the preference for the claim that the quality of the production is sufficient became well supported.

5.3 Continuous Distributions: Transformations and Mixtures

The calculus presented in the previous sections and in Chaps. 2 and 4 is difficult to adapt to most recognised classes of continuous distributions. The parameters of some distributions are not directly the targets of inference. For example, a parameter of the beta or gamma distribution is rarely a suitable object for a decision or for another form of inference. The expectations of these distributions are nonlinear functions of their parameters. These classes of distributions are not closed with respect to linear transformations. For example, a shift (adding a constant) would be readily recognised on a beta or gamma distribution by its changed support, e.g., from $(0, 1)$ to $(s, 1+s)$.

Monotone transformations and finite mixtures are two general approaches for extending the methods for normal and t to wider classes of distributions. We do not have to dwell on the extension by transformations, because it is usually more practical to work on the scale on which normality applies. With a nonlinear transformation we lose the convenient link of the parameters with the moments of the distribution. Recall, as an example, that the expectation and variance of the lognormal distribution, the exponential of $\mathcal{N}(\mu, \sigma^2)$, are $\exp(\mu+\frac{1}{2}\sigma^2)$ and $\exp(2\mu+\sigma^2)\{\exp(\sigma^2)-1\}$. These are rather complex functions of the log-scale mean μ and variance σ^2, as are their inverses. Also, we are more comfortable working with symmetric distributions.

Another way of extending the methods presented thus far is by considering mixtures of distributions. A finite mixture of distributions with densities $f_1, \ldots, f_K$ and probabilities $\pi_1, \ldots, \pi_K$ that add up to unity is defined by the density

$$f(x) = \pi_1 f_1(x) + \cdots + \pi_K f_K(x), \tag{5.6}$$

that is, as a composition (a linear convex combination) of the component densities f_k McLachlan and Peel 2000. Mixtures are easy to work with because the principal operations of expectation and differentiation with mixture densities (or probabilities) are no more difficult than the same operations with the component densities.

A draw from a finite mixture can be generated in the following two stages. First we draw the component k from the multinomial distribution $\mathcal{K}$ with probabilities $\pi_1, \ldots, \pi_K$. Then, if the draw is k, we draw from the distribution (component) with density f_k. This random variable is denoted by $X_{\mathcal{K}}$. More general mixtures are defined by a mixing distribution $\mathcal{M}$ with support in a set S and a set of distributions $\mathcal{H} = \{\mathcal{H}_s\}$ with index $s \in S$. When the mixing distribution is defined on the integers $0, 1, \ldots, \mathcal{H}$ is a sequence of distributions.

Finite mixtures are fitted by the EM algorithm (Dempster et al. 1977). The EM algorithm is a general approach to maximum likelihood estimation suitable when direct maximisation (e.g., by the Fisher scoring algorithm) is very difficult or not feasible. That is the case with finite mixtures, for which the likelihood is a product of densities of the form (5.6). To implement the EM algorithm, we declare first what we regard as the missing data. A purposeful way of doing this is by supplementing (augmenting) the recorded data so that the hypothetical complete (augmented) dataset would be easy to analyse. For mixtures, the identity of the component to which each observation belongs is the obvious choice.

The EM algorithm proceeds by iterations. Each iteration comprises step E, in which the contribution to the log-likelihood that involves the missing data is estimated, and step M, in which the method intended for the complete data is applied. For finite mixtures, the E step entails evaluation of the conditional probabilities of each observational unit $i = 1, \ldots, n$ belonging to the mixture components $k = 1, \ldots, K$, given the current estimates;

$$\hat{r}_{ik} = \frac{\hat{\pi}_k \hat{f}_k(x_i)}{\hat{\pi}_1 \hat{f}_1(x_i) + \cdots + \hat{\pi}_K \hat{f}_K(x_i)},$$

where the circumflex ^ used on the right-hand side indicates (provisional) estimation by the preceding M step. In the M step, the procedure that would have been appropriate with the complete data is applied, with the missing items replaced by their conditional expectations evaluated in the preceding E step. For finite mixtures, this is the weighted version of the complete-data procedure for each component k, with the weights set to $\hat{r}_{ik}$. See Longford and D'Urso (2011) for details and an application in which one component is intended specifically for outliers.

The expectation of a mixture is equal to the combination of the expectations μ_k of the basis distributions, $\bar{\mu} = \pi_1 \mu_1 + \cdots + \pi_K \mu_K$, but for the variance we have a more complex expression,

$$\mathrm{var}(X) = \sum_{k=1}^{K} \pi_k \sigma_k^2 + \sum_{k=1}^{K} \pi_k (\mu_k - \bar{\mu})^2, \tag{5.7}$$

where σ_k^2 is the variance of component k. Of course, this expression is valid only when every variance σ_k^2 is finite. We can approximate any distribution by a finite mixture of normals, although some other classes of distributions, the uniform and the degenerate in particular, also have this property of being dense in the space of all distributions. The following identity summarises how easy it is to evaluate the expected loss with mixtures of distributions.

$$Q = \int L(\hat{\theta}, \theta) \sum_{k=1}^{K} \pi_k f_k(\theta) \, d\theta = \sum_{k=1}^{K} \pi_k \int L(\hat{\theta}, \theta) f_k(\theta) \, d\theta = \sum_{k=1}^{K} \pi_k Q^{(k)},$$
(5.8)

where $Q^{(k)}$ is the expected loss with mixture component k.

Further, the derivative of Q is equal to the π_k-linear combination of the derivatives of $Q^{(k)}$, so minimisation of the expected loss by the Newton-Raphson algorithm is not any more complex than the corresponding task for a component distribution.

A linear combination of loss functions for the same target θ, $L(\hat{\theta}, \theta) = \zeta_1 L_1(\hat{\theta}, \theta) + \cdots + \zeta_H L_H(\hat{\theta}, \theta)$, with positive coefficients ζ_h, is also a loss function. An identity similar to (5.8) applies for such a composite loss function L. Thus, the variety of the loss functions that can be handled is expanded substantially. Note, however, that the number of terms to be evaluated also expands, to $H \times K$, so combinations with many terms for the loss (H) or the density (K) raise the complexity of the evaluations.

5.4 Problems, Exercises and Suggested Reading

1. Review the properties of the Poisson distribution, including its no-memory property and closure with respect to convolution. Prove that the binomial distribution converges in probability to the Poisson as $n \to +\infty$ if the probabilities $p = p_n$ are such that np_n has a finite positive limit.
2. Prove the expressions for the expectation and variance of a finite mixture.
3. Implement a process for sampling from a finite mixture distribution on the computer. Verify empirically the expressions for the expectation and variance of a finite mixture.
4. Implement the EM algorithm for a mixture of a small number of normal distributions. Adapt it to fitting a mixture of uniform distributions and relate the fit to a histogram. Generate a dataset from a mixture of normal distributions, with at least 1,000 observations. Assess how well the components are recovered by the fit when you set their number correctly, incorrectly, and when you use a data-based criterion for setting it.
5. Suggested reading about the lognormal distribution: Longford (2009a) and the references therein. Discuss the difficulties with inferences about the mean of a lognormal distribution with an asymmetric loss.
6. Explore by simulations the Poisson mixture of Poisson distributions. That is, a draw is made from a Poisson distribution and its value is the parameter for an independent draw from another Poisson distribution. Compare this mixture distribution with the distribution defined as the sum of independent and identically distributed Poisson variables, the number of which itself has a Poisson distribution.

7. A project: Description of mixtures of Poisson distributions. When is a mixture of two Poissons bimodal? Could a mixture of two (or more) Poissons have a Poisson distribution? Let X and Y be independent variables, both with a Poisson distribution. Study the properties of the interweaved Poisson, defined as a mixture of $2X$ and $2Y - 1$. Study the problem of deciding which of the Poisson distributions, the even (X) or the odd (Y) has a greater expectation.

8. Suggested reading about the problem of estimating a probability based on a binomial trial with no successes in n attempts: Winkler et al. (2002), Longford (2009b, 2010).

9. Find out about the Pareto class of distributions from Arnold (1983), Embrechts and Schmidli (1994) or other references. Formulate problems in terms of deciding between two courses of action depending on the value of the parameter of the distribution.

10. Discuss the relevance of decision theory to modelling extreme events, such as maximum level of a river, incidents of high pollution, failures of safety in public transport, natural disasters (earthquakes and volcanic activity) and related planning (investment). Suggested reading: Coles (2001).

References

Arnold, B. C. (1983). *Pareto distributions*. Burtonsville, MD: International Co-operative Publishing House.

Coles, S. (2001). *An introduction to statistical modelling of extreme events*. London: Springer.

Dempster, A. P., Laird, N. M., & Rubin, D. B. (1977). Maximum likelihood for incomplete data via the EM algorithm. *Journal of the Royal Statistical Society: Series B, 39*, 1–38.

Embrechts, P., & Schmidli, H. (1994). *Modelling of extremal events in insurance and finance*. New York: Springer.

Longford, N. T. (2009a). Inference with the lognormal distribution. *Journal of Statistical Planning and Inference, 139*, 2329–2340.

Longford, N. T. (2009b). Analysis of all-zero binomial outcomes. *Journal of Applied Statistics, 36*, 1259–1265.

Longford, N. T. (2010). Bayesian decision making about small binomial rates with uncertainty about the prior. *The American Statistician, 64*, 164–169.

Longford, N. T., & D'Urso, P. (2011). Mixture models with an improper component. *Journal of Applied Statistics, 38*, 2511–2521.

McLachlan, G. J., & Peel, D. (2000). *Finite mixture models*. New York: Wiley.

Winkler, R. L., Smith, J. E., & Fryback, D. G. (2002). The role of informative priors in zero-numerator problems: Being conservative versus being candid. *American Statistician, 56*, 1–4.

Chapter 6
Classification

Classification is a term used for the process of assigning an observed unit to one of a small number of labelled groups (classes), such as 'ordinary' and 'unusual or 'positive', 'neutral' and 'negative'. The groups are exclusive and exhaustive—a single group is appropriate for every unit. Common applications of classification arise in medical screening, educational tests and licencing examinations, fraud detection and when searching for units with exceptional attributes. The groups may be well defined a priori, or their definition is based on the analysis of a collection of observed units. The term 'misclassification' is used for assigning a unit to an inappropriate group, a group to which the unit does not belong. We deal with the setting of two groups of units, called positives and negatives, in which there are two kinds of inappropriate assignments; the corresponding (misclassified) units are called false positives and false negatives. Our task is to minimise the expected loss associated with such misclassification.

6.1 Introduction

As an example, we consider the process of screening for the precursors of a serious medical condition, such as a chronic debilitating disease that manifests itself at an old age. Subjects are screened in middle age, when some effective though expensive and inconvenient measures can be taken to reduce the severity and slow down the progression of the condition in the future. After the onset of the condition, there is no effective treatment, merely a passive management of the disease. There are obvious advantages to identifying future cases of the disease, but treating many candidates (strongly preferring to err on the side of false positiveness) would overwhelm the healthcare system.

We assume that a single variable Y is observed on a sequence of unrelated units (subjects). This variable, called the *marker*, is related to the classification of the units as positives and negatives. Say, positives tend to have large and negatives small values of Y. We want to set a threshold T so that every unit with $Y > T$ would be classified as

N. T. Longford, *Statistical Decision Theory*,
SpringerBriefs in Statistics, DOI: 10.1007/978-3-642-40433-7_6,
© The Author(s) 2013

positive, and every other unit as negative. As the introduction indicates, errors of the two kinds (declaring false positives and false negatives) have unequal consequences. In our example, a false positive is associated with unnecessary anguish, making inappropriate plans for the future, administration of treatments and procedures that are inappropriate or unnecessary, and the like. A false negative is associated with much more grave consequences—a medical condition that could have been treated, but by the time its symptoms become transparent it is too late.

As a contrasting example, suppose we are looking for suitable candidates for a set of identical jobs; there are many applicants for the few vacancies, we are certain that many of them are suitable, but our interviewing process and selection are imperfect and we do not reliably classify the applicants as suitable and unsuitable. In this case, the employer wants suitable workers in the first instance, so he prefers to commit errors of failing to identify positives—a false positive has more severe consequences than a false negative. Of course, the labels 'positive' and 'negative' can be switched, so it is not an essential feature of the example that one or the other kind of error has more severe consequences; the key feature is *how* much more severe they are for one than for the other kind of error.

6.2 Normally Distributed Marker

We are dealing with two variables, a marker Y and a (correct) classification U, defined in a population of units. In a typical example from medical screening, the value of Y can be established for any person willing to undergo a routine examination, whereas the value of U is revealed only with the passing of time. We assume first that Y has normal conditional distributions within the groups $U = 1$ (positives) and $U = 0$ (negatives), $\mathcal{N}(\mu_1, \sigma_1^2)$ and $\mathcal{N}(\mu_0, \sigma_0^2)$, respectively. We want to set a threshold T, so that every unit with $Y > T$ would be declared as positive, and every other as negative.

This problem has a long history in statistics, with a seminal contribution made by Youden (1950) who proposed to choose T for which the quantity

$$P(U = 1 \mid Y > T) + P(U = 0 \mid Y < T),$$

called the Youden index, is maximised. These probabilities, called respectively sensitivity and specificity, refer to appropriate decisions, but the index can also be expressed in terms of conditional probabilities of the two kinds of misclassification, and the condition to minimise their total. As an adaptation, their weighted total could be minimised. Weights are attached to the probabilities to reflect the uneven gravity of the two kinds of error.

Instead of the established approach, we prefer to work with the conditional probabilities $P(Y > T \mid U = 0)$ and $P(Y < T \mid U = 1)$, that is, to condition on the (future) disease status U. The two probababilities correspond to the two kinds of error. Suppose a false positive, for which $Y > T$ but $U = 0$, is associated with loss $L_+(Y; T) = (Y - T)^2$ and a false negative, for which $Y < T$ but $U = 1$, with loss

$L_-(Y; T) = R(Y - T)^2$. For appropriate assignments, when $Y > T$ and $U = 1$, and when $Y < T$ and $U = 0$, no loss is incurred. The loss function is $L = L_- + L_+$, but at least one of the contributions vanishes, so $L = L_+$ or $L = L_-$ for every unit; also $L = \max(L_+, L_-)$. Let $p_1 = P(U = 1)$ be the prevalence of positives, and denote the prevalence of negatives by $p_0 = 1 - p_1$.

The expected loss, a function of the threshold T, is

$$Q(T) = \frac{p_0}{\sigma_0} \int_T^{+\infty} (y - T)^2 \, \phi\left(\frac{y - \mu_0}{\sigma_0}\right) dy + \frac{p_1 R}{\sigma_1} \int_{-\infty}^T (y - T)^2 \, \phi\left(\frac{y - \mu_1}{\sigma_1}\right) dy.$$

We encountered similar expressions in Chap. 2, e.g., in Eqs. (2.1) and (2.2), so we leave for an exercise (or revision) the derivation of the expression

$$Q(T) = p_0 \sigma_0^2 \left[\left(1 + Z_0^2\right) \{1 - \Phi(Z_0)\} - Z_0 \phi(Z_0)\right]$$
$$+ p_1 R \sigma_1^2 \left[\left(1 + Z_1^2\right) \{1 - \Phi(Z_1)\} - Z_1 \phi(Z_1)\right],$$

where $Z_0 = (T - \mu_0)/\sigma_0$ and $Z_1 = (\mu_1 - T)/\sigma_1$. The minimum of Q is found by the Newton-Raphson algorithm; the derivatives required for it are

$$\frac{\partial Q}{\partial T} = 2p_0 \sigma_0 \left[Z_0 \{1 - \Phi(Z_0)\} - \phi(Z_0)\right]$$
$$- 2p_1 R \sigma_1 \left[Z_1 \{1 - \Phi(Z_1)\} - \phi(Z_1)\right]$$
$$\frac{\partial^2 Q}{\partial T^2} = 2p_0 \{1 - \Phi(Z_0)\} + 2p_1 R \{1 - \Phi(Z_1)\}.$$

The second-order derivative is positive and Q diverges to $+\infty$ for very large and very small (negative) T. Therefore, Q has its unique minimum at the root of $\partial Q/\partial T$. Obvious initial values for the Newton-Raphson iterations are $T^{(0)} = \frac{1}{2}(\mu_0 + \mu_1)$ or the mean, median or p_0-quantile of the observations Y.

An alternative to the piecewise quadratic loss function is the piecewise linear. It is defined as $Y - T$ for false positives and as $R(T - Y)$ for false negatives. The expected loss and its derivatives are

$$Q(T) = \frac{p_0}{\sigma_0} \int_T^{+\infty} (y - T) \phi\left(\frac{y - \mu_0}{\sigma_0}\right) dy + \frac{p_1 R}{\sigma_1} \int_{-\infty}^T (T - y) \phi\left(\frac{y - \mu_1}{\sigma_1}\right) dy$$
$$= -p_0 \sigma_0 \left[Z_0 \{1 - \Phi(Z_0)\} - \phi(Z_0)\right] - p_1 R \sigma_1 \left[Z_1 \{1 - \Phi(Z_1)\} - \phi(Z_1)\right]$$
$$\frac{\partial Q}{\partial T} = -p_0 \{1 - \Phi(Z_0)\} + p_1 R \{1 - \Phi(Z_1)\}$$
$$\frac{\partial^2 Q}{\partial T^2} = \frac{p_0}{\sigma_0} \phi(Z_0) + \frac{p_1 R}{\sigma_1} \phi(Z_1).$$

For the piecewise constant loss, defined as unity for false positives and R for false negatives, the corresponding identities are

$$Q(T) = p_0 \{1 - \Phi(Z_0)\} + p_1 R \{1 - \Phi(Z_1)\}$$

$$\frac{\partial Q}{\partial T} = -\frac{p_0}{\sigma_0} \phi(Z_0) + \frac{p_1 R}{\sigma_1} \phi(Z_1)$$

$$\frac{\partial^2 Q}{\partial T^2} = \frac{p_0}{\sigma_0^2} Z_0 \phi(Z_0) + \frac{p_1 R}{\sigma_1^2} Z_1 \phi(Z_1).$$

With either of the three sets of loss functions, we lose no generality by assuming that $\mu_0 = 0$ and $\sigma_0^2 = 1$, because a linear transformation of the marker Y does not alter the nature of the problem. Further, the optimal threshold depends on p_1 and R only through the penalised odds ratio $\rho = p_1 R / p_0$. Thus, we can describe the entire set of solutions as a function of $\mu_1^* = (\mu_1 - \mu_0)/\sigma_0$, $\sigma_1^* = \sigma_1/\sigma_0$ and ρ. Owing to symmetry of the normal distribution, we can reduce our attention to $\mu_1^* > 0$. Further, we can disregard settings with rare positives (small p_1) and small R (small values of ρ) and rare negatives (small p_0) and large R (large values of ρ), because they represent unrealistic situations. Settings with small μ_1^* and large σ_1^* are also unrealistic because classification in them is full of errors; the marker is ineffective.

Figure 6.1 presents continua of solutions for the three kernels, a set of variances indicated at the margins and values of ρ set to 2.0 (curves drawn in black) and $\rho = 0.5$ (gray). For the linear and quadratic kernels, greater standardised mean μ_1^* and smaller relative variance σ_1^{2*} are associated with higher threshold T. The threshold is higher and the expected loss is lower for greater ρ. For the quadratic kernel, the threshold converges to the value $T_\infty \doteq 0.611$ as $\mu_1^* \to +\infty$ for all values of σ_1^{2*} and ρ. The expected loss as a function of T converges to $L_\infty \doteq 0.169$ as $\mu_1^* \to +\infty$.

For the absolute kernel and $\rho = 0.5$, the threshold functions intersect at around $\mu_1 = 2.4$ and the expected losses at around $\mu_1^* = 1.05$. As functions of σ_1^{2*}, they are increasing up to that point, and they decrease for $\mu_1^* > 1.05$.

The dependence of T on the values of μ_1^*, σ_1^{2*} and ρ can be explored with the aid of the implicit function theorem (IFT). For the quadratic kernel, let $F = \frac{1}{2} p_0^{-1} \partial Q / \partial T$, as a function of T, μ_1^*, σ_1^{2*} and ρ. According to IFT,

$$\frac{\partial T}{\partial \mu_1^*} = -\frac{\partial F}{\partial \mu_1^*} \Big/ \frac{\partial F}{\partial T},$$

if any two of these derivatives exist, and $\partial F / \partial T \neq 0$. By simple operations we obtain for the piecewise quadratic loss the identities

$$\frac{\partial F}{\partial T} = 1 - \Phi(T) + \rho \{1 - \Phi(Z_1)\}$$

$$\frac{\partial F}{\partial \mu_1^*} = -\rho \{1 - \Phi(Z_1)\}$$

$$\frac{\partial F}{\partial \sigma_1^*} = \rho \phi(Z_1)$$

$$\frac{\partial F}{\partial \rho} = -\sigma_1^* [Z_1 \{1 - \Phi(Z_1)\} - \phi(Z_1)].$$

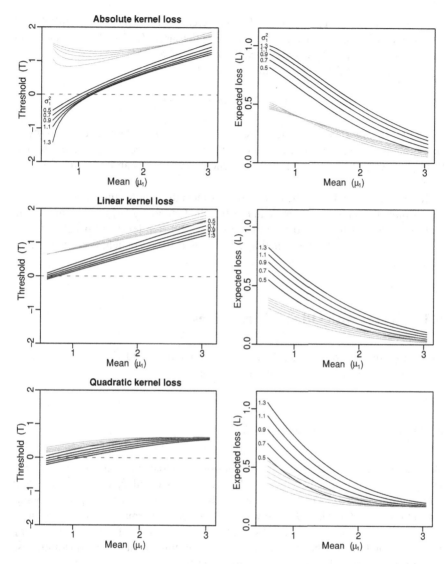

Fig. 6.1 Classification thresholds for the three kernels as functions of the standardised mean difference μ_1^* for variances σ_1^{2*} (indicated at the margins) and $\rho = 2.0$ (*black*) and $\rho = 0.5$ (*gray*)

Hence

$$\frac{\partial T}{\partial \mu_1^*} = \frac{\rho\{1 - \Phi(Z_1)\}}{1 - \Phi(T) + \rho\{1 - \Phi(Z_1)\}},$$

and since this is positive throughout, T is an increasing function of μ_1^*. Similarly, we conclude that T is a decreasing function of σ_1^*. For large Z_1, $1 - \Phi(Z_1)$ and $\phi(Z_1)$ are very small, so F depends very weakly on both μ_1^* and σ_1^*.

To show that $\partial F / \partial \rho < 0$, we study the function

$$g(z) = z\{1 - \Phi(z)\} - \phi(z),$$

related to the brackets, [ ], in the expression for $\partial F / \partial \rho$. Its derivative is $1 - \Phi(z)$, so g is an increasing function. The limits of $g(z)$ as $z \to -\infty$ and $+\infty$ are $-\infty$ and 0, respectively. Therefore $g(z) < 0$ throughout, and so T is an increasing function of ρ.

It is rather ironic that the behaviour of T is most difficult to study analytically for the absolute kernel. In the example in Fig. 6.1, the functions $T(\sigma_1^*)$ are not monotone and they intersect for $\rho = 0.5$, but not for $\rho = 2.0$ (top left-hand panel), and for $\rho = 2.0$ and $\sigma_1^{2*} = 1.3$ the function $T(\mu_1^*)$ has a steep gradient for small values of μ_1^*.

6.3 Markers with Other Distributions

In this section, we solve the problem of setting the threshold T for a few common alternatives to the normal distribution of the marker.

6.3.1 Markers with t Distribution

We assume that the conditional distributions of the marker Y, given the group $U = 0$ or 1, are scaled t; that is, $(Y - \mu_u)/\sigma_u$ has t distribution with k_u degrees of freedom within group $u = 0, 1$. The derivations of the expressions for the expected loss as a function of the threshold T follow the outline for the normal distribution. Denote by ψ_k the density of the central t distribution with k degrees of freedom, and by Ψ_k the corresponding distribution function. For integrating the linear and quadratic loss with respect to the t density, we apply the identity

$$\int x\psi_k(x)\,dx = -\frac{1}{\gamma_k}\psi_{k-2}(\gamma_k x),$$

where $\gamma_k = \sqrt{1 - 2/k}$ and $k > 2$; see (4.11). By letting $k \to +\infty$ we obtain the counterpart of this identity for the normal distribution, $\int x\phi(x)\,dx = -\phi(x)$.

The false negatives contribute to the expected piecewise quadratic loss $Q = Q_+ + Q_-$ by

$$Q_+ = \frac{p_0}{\sigma_0} \int_T^{+\infty} (y - T)^2 \, \psi_{k_0}\left(\frac{y - \mu_0}{\sigma_0}\right) dy$$

$$= p_0 \sigma_0^2 \int_{Z_0}^{+\infty} (z - Z_0)^2 \, \psi_{k_0}(z) \, dz$$

$$= p_0 \sigma_0^2 \, G_2(k_0, Z_0),$$

where $Z_0 = (T - \mu_0)/\sigma_0$ and

$$G_2(k, Z) = Z^2 \{1 - \Psi_k(Z)\} - \frac{Z}{\gamma_k} \psi_{k-2}(\gamma_k Z) + \frac{1}{\gamma_k^2} \{1 - \Psi_{k-2}(\gamma_k Z)\}.$$

By similar steps we obtain the contribution made by the false positives,

$$Q_- = p_1 R \sigma_1^2 \, G_2(k_1, Z_1),$$

where $Z_1 = (\mu_1 - T)/\sigma_1$. The derivatives of the expected loss Q are

$$\frac{\partial Q}{\partial T} = 2 \{ p_0 \sigma_0 \, G_1(k_0, Z_0) - p_1 R \sigma_1 G_1(k_1, Z_1) \},$$

where

$$G_1(k, Z) = Z \{1 - \Psi_k(Z)\} - \frac{1}{\gamma_k} \psi_{k-2}(\gamma_k Z).$$

and

$$\frac{\partial^2 Q}{\partial T^2} = 2 \{ p_0 \, G_0(k_0, Z_0) + p_1 R G_0(k_1, Z_1) \},$$

with $G_0(k, Z) = 1 - \Psi_k(Z)$.

For the piecewise linear loss, the following identities apply:

$$Q = - p_0 \sigma_0 \, G_1(k_0, Z_0) - p_1 R \sigma_1 G_1(k_1, Z_1)$$

$$\frac{\partial Q}{\partial T} = - p_0 \, G_0(k_0, Z_0) + p_1 R G_0(k_1, Z_1)$$

$$\frac{\partial^2 Q}{\partial T^2} = \frac{p_0}{\sigma_0} \psi_{k_0}(Z_0) + \frac{p_1 R}{\sigma_1} \psi_{k_1}(Z_1),$$

and for the piecewise constant loss, we have

$$Q = p_0 \, G_0(k_0, Z_0) + p_1 R G_0(k_1, Z_1)$$

$$\frac{\partial Q}{\partial T} = - \frac{p_0}{\sigma_0} \psi_{k_0}(Z_0) + \frac{p_1 R}{\sigma_1} \psi_{k_1}(Z_1)$$

$$\frac{\partial^2 Q}{\partial T^2} = \frac{p_0 Z_0}{\sigma_0^2} \psi_{k_0+2}\left(\frac{Z_0}{\gamma_{k_0+2}}\right) + \frac{p_1 Z_1}{\sigma_1^2} \psi_{k_1+2}\left(\frac{Z_1}{\gamma_{k_1+2}}\right).$$

For k_0 and k_1 diverging to infinity, these expressions converge to the corresponding expressions for the normally distributed marker.

6.3.2 Beta Distributed Markers

Some markers are defined on a scale with limits at either extreme, such as $(0,1)$ or $(0, 100)$. For such variables the beta distribution, or its linear transformation, is the first choice one would contemplate. The beta distributions are given by the densities

$$\xi(x; \alpha, \beta) = \frac{\Gamma(\alpha + \beta)}{\Gamma(\alpha)\,\Gamma(\beta)}\, x^{\alpha-1}(1-x)^{\beta-1},$$

where $\alpha > 0$ and $\beta > 0$ are parameters and $x \in (0, 1)$. Their expectations are $\lambda = \alpha/(\alpha + \beta)$ and variances $\lambda(1 - \lambda)/(\alpha + \beta + 1)$. The essential element of the calculus for the integrals involving beta densities is the identity

$$x\xi(x; \alpha, \beta) = \lambda\xi(x; \alpha + 1, \beta). \tag{6.1}$$

Suppose the parameters associated with the two groups are (α_u, β_u), $u = 0, 1$, and denote $\lambda_u = \alpha_u/(\alpha_u + \beta_u)$ and $\tau_u = \lambda_u (\alpha_u + 1)/(\alpha_u + \beta_u + 1)$. The contribution Q_+ of the false negatives to the expected loss is

$$
\begin{aligned}
Q_+ &= p_0 \int_T^1 (y - T)^2\, \xi(y; \alpha_0, \beta_0)\, dy \\
&= p_0\tau_0 \int_T^1 \xi(y; \alpha_0 + 2, \beta_0)\, dy - 2p_0 T \lambda_0 \int_T^1 \xi(y; \alpha_0 + 1, \beta_0)\, dy \\
&\quad + p_0 T^2 \int_T^1 \xi(y; \alpha_0, \beta_0)\, dy \\
&= p_0 \left\{ \frac{\lambda_0(1 - \lambda_0)}{\alpha_0 + \beta_0 + 1} + (T - \lambda_0)^2 \right\} - p_0 H_2(T; \lambda_0, \tau_0, \alpha_0, \beta_0),
\end{aligned}
$$

where

$$H_2(T; \lambda, \tau, \alpha, \beta) = \tau B(T; \alpha + 2, \beta) - 2T\lambda B(T; \alpha + 1, \beta) + T^2 B(T; \alpha, \beta),$$

and B is the beta distribution function. The first term in the concluding expression for Q_+ is the p_0-multiple of the mean squared deviation of the marker around the threshold. By similar steps we obtain the complementary expression

$$Q_- = p_1 R H_2(T; \lambda_1, \tau_1, \alpha_1, \beta_1).$$

For piecewise linear loss,

$$Q_+ = p_0\,(\lambda_0 - T) - p_0\,H_1\,(T;\lambda_0,\alpha_0,\beta_0)$$
$$Q_- = -p_1 R H_1\,(T;\lambda_0,\alpha_0,\beta_0),$$

where

$$H_1(T;\lambda,\alpha,\beta) = \lambda B(T;\alpha+1,\beta) - T B(T;\alpha,\beta).$$

For piecewise constant loss,

$$Q = Q_- + Q_+ = p_0\,\{1 - B(T;\alpha_0,\beta_0)\} + p_1 R B(T;\alpha_1,\beta_1).$$

Except for the piecewise constant loss, the expressions for the first- and second-order derivatives of the expected loss Q are easy to derive, but they are lengthy and it is more practical to approximate the derivatives numerically.

6.3.3 Gamma Distributed Markers

For the class of gamma distributions, given by the densities

$$g(x;\nu,s) = \frac{1}{\Gamma(s)}\,\nu^s x^{s-1}\,\exp(-x\nu),$$

we have the identity

$$x g(x;\nu,s) = \frac{s}{\nu}\,g(x;\nu,s+1),$$

similar to (6.1). Denote by G the distribution function of gamma and by ν_h and s_h the parameters of the conditional gamma distributions of the marker Y given the class $U = h, h = 0, 1$.

For piecewise quadratic loss,

$$Q(T) = p_0\left\{\frac{s_0}{\nu_0^2} + \left(T - \frac{s_0}{\nu_0}\right)^2\right\} - p_0 G_2\,(T;\nu_0,s_0) + p_1 R G_2\,(T;\nu_1,s_1),$$

where

$$G_2(T;\nu,s) = \frac{s(s+1)}{\nu^2}\,G(T;\nu,s+2) - \frac{2sT}{\nu}\,G(T;\nu,s+1) + T^2 G(T;\nu,s).$$

For piecewise linear loss,

$$Q(T) = p_0\left(\frac{s_0}{\nu_0} - T\right) - p_0 G_1(T;\nu_0,s_0) - p_1 G_1(T;\nu_1,s_1),$$

where

$$G_1(T; \nu, s) = \frac{s}{\nu} G(T; \nu, s + 1) - T G(T; \nu, s).$$

For piecewise constant loss,

$$Q(T) = p_0 \{1 - G(T; \nu_0, s_0)\} + p_1 R G(T; \nu_1, s_1).$$

Except for the piecewise linear loss, tedious analytical evaluation of the derivative $\partial Q / \partial T$ is best replaced by its numerical approximation $\{Q(T + h) - Q(T)\}/h$ for suitable small h.

6.4 Looking for Contaminants

In the following application, we use the qualifiers ordinary and exceptional instead of negative and positive, to avoid confusion with the actual values of the marker. Suppose the marker is distributed according to $\mathcal{N}(0, 1)$ for ordinary units, and exceptional units have either positive or negative values distant from zero. In an example, their distribution is a $\frac{1}{2} - \frac{1}{2}$ mixture of $\mathcal{N}(-3, 2)$ and $\mathcal{N}(4, 3)$. The probability of a unit being ordinary is p_0, and the probabilities of the two components for the exceptional units are p_{1L} and p_{1U}; in our example, $p_{1L} = p_{1U} = 0.025$. The densities, scaled by their probabilities, are drawn in Fig. 6.2, together with the values of the marker for a set of 250 units which we wish to classify. The observed values are marked at the top of the diagram (the exceptional units have longer ticks), and those misclassified with

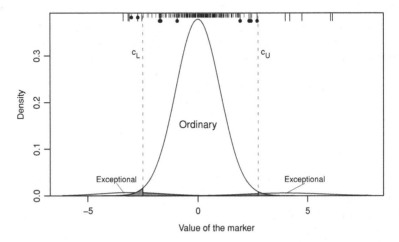

Fig. 6.2 Example with ordinary and exceptional units, with the setting $\mu_0 = 0$, $\sigma_0^2 = 1$, $\mu_{1L} = -3$, $\sigma_{1L}^2 = 2$, $\mu_{1U} = 4$, $\sigma_{1U}^2 = 3$, $p_0 = 0.95$, $p_{1L} = p_{1U} = 0.025$. The interval (c_L, c_U), marked by *vertical dashes*, is $(-2.5, 2.75)$

respect to the thresholds indicated by the vertical dashes have a black disk attached. The areas under the densities that correspond to misclassification are highlighted by shading. With real data, we would not know which unit is ordinary and which is exceptional.

Our task is to define an interval (c_L, c_U) such that a unit with value of the marker in it would be classified as ordinary, and a unit outside the interval as exceptional. Losses are incurred when the marker of an exceptional unit falls into this interval and when the marker of an ordinary unit is outside. Thus, the expected loss with the piecewise constant loss function with penalty ratio R (affecting units incorrectly declared as ordinary) is

$$
Q = p_0 \left\{ 1 - \frac{1}{\sigma_0} \int_{c_L}^{c_U} \phi\left(\frac{x - \mu_0}{\sigma_0}\right) dx \right\}
$$
$$
+ R \frac{p_{1L}}{\sigma_{1L}} \int_{c_L}^{c_U} \phi\left(\frac{x - \mu_{1L}}{\sigma_{1L}}\right) dx + R \frac{p_{1U}}{\sigma_{1U}} \int_{c_L}^{c_U} \phi\left(\frac{x - \mu_{1U}}{\sigma_{1U}}\right) dx \,;
$$

the distributions involved are $\mathcal{N}(\mu_0, \sigma_0^2)$ for the ordinary units, $\mathcal{N}(\mu_{1L}, \sigma_{1L}^2)$ for the 'negative' exceptions and $\mathcal{N}(\mu_{1U}, \sigma_{1U}^2)$ for the 'positive' exceptions. We assume that σ_{1L}^2 and σ_{1U}^2 are both greater than σ_0^2.

Standard operations lead to the expression for the expected loss in terms of the normal distribution function Φ:

$$
Q(c_L, c_U; R) = p_0 \left\{ 1 - \Phi(d_{U0}) + \Phi(d_{L0}) \right\}
$$
$$
+ R \left[p_{1L} \left\{ \Phi(d_{U1L}) - \Phi(d_{L1L}) \right\} + p_{1U} \left\{ \Phi(d_{U1U}) - \Phi(d_{L1U}) \right\} \right],
$$

where $d_{L1U} = (c_L - \mu_{1U})/\sigma_{1U}$, and similarly for the other quantities d. The arguments c_L and c_U appear in distinct sets of terms, so the expected loss can be minimised for them separately. We have

$$
\frac{\partial Q}{\partial c_L} = f(c_L; R)
$$
$$
\frac{\partial Q}{\partial c_U} = -f(c_U; R),
$$

where

$$
f(c; R) = \frac{p_0}{\sigma_0} \phi\left(\frac{c - \mu_0}{\sigma_0}\right) - \frac{p_{1L}R}{\sigma_{1L}} \phi\left(\frac{c - \mu_{1L}}{\sigma_{1L}}\right) - \frac{p_{1U}R}{\sigma_{1U}} \phi\left(\frac{c - \mu_{1U}}{\sigma_{1U}}\right). \quad (6.2)
$$

Thus, minimisation of L coincides with the problem of finding the roots of $f(c, R)$ for c. Although the function f depends also on the probabilities p, means μ and variances σ^2, we emphasise only the dependence on R, because that is often the focus of a sensitivity analysis. For example, instead of a single value of R a plausible range of values, (R_L, R_U), is declared. The nature of the problem is not altered if we standardise the marker so that its conditional mean and variance among the ordinary units are zero and unity, respectively. Further, the dependence on the probabilities can be reduced to the dependence on the ratios $\rho_L = p_{1L}/p_0$ and $\rho_U = p_{1U}/p_0$, with the obvious simplification when $p_{1L} = p_{1U}$. Also, only the ratios σ_{1L}/σ_0 and σ_{1U}/σ_0 matter.

As $c \rightarrow \pm\infty$, $f(c; R)$ converges to zero for every R. For c in the vicinity of $\pm\infty$, $f(c, R)$ is negative, because the dominant contribution to f in (6.2) is from the density with the largest variance, which we assume to be either σ_{1L}^2 or σ_{1U}^2. Therefore $f(c; R)$ has at least two roots when it is positive for some c. For sufficiently large values of $R\rho_U$ and $R\rho_L$, $f(c; R)$ is negative throughout. In that case, every unit is classified as exceptional, because the penalty for declaring a false negative is prohibitive. The roots of f are found by the Newton-Raphson algorithm. The distinct roots are found by different settings of the initial solution, close to either of the means μ_{1L} and μ_{1U}.

Example

The continuum of the optimal thresholds for the setting of the example in Fig. 6.2 is presented in Fig. 6.3 for the penalty ratios R in the range $(1.0, 349.1)$, on the linear and log scales. With increasing R the range of values of the marker that correspond to the classification 'ordinary' is shrinking until around $R_\dagger = 349.1$, where the two threshold functions meet. At that point, $c_L(R_\dagger) = c_U(R_\dagger) \doteq 0.20$. For $R > R_\dagger$, $\partial f/\partial c$ is negative throughout and has no root. Every unit is then classified as a

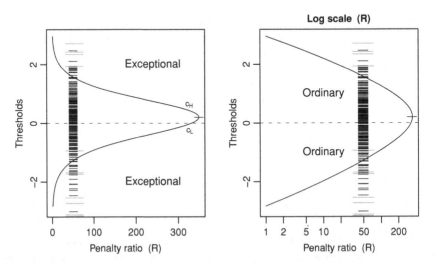

Fig. 6.3 The optimal thresholds for $R \in (1, 349)$ for the setting of Fig. 6.2. The (simulated) data is indicated by the horizontal segments, *gray* coloured and extended for the exceptional units. The words 'Ordinary' and 'Exceptional' indicate assignment of the units, not their status U

contaminant. The observations are indicated by horizontal segments drawn in the range $R \in (40, 60)$; the segments for the exceptional units are extended and drawn in gray colour. Note that some observations are off the vertical scale.

Suppose the plausible range of R is $(40, 60)$. Then every horizontal segment that intersects the function c_L or c_U in this range has a different classification at $R = 40$ and $R = 60$—its classification is equivocal. There are 11 such units (out of 250), all of them ordinary; their segments intersect c_L in five and c_U in six instances. One exceptional unit and 26 ordinary units $(16 + 10)$ are misclassified for all $R \in (40, 60)$.

The functions $c_L(R)$ and $c_U(R)$ are evaluated by the Newton-Raphson algorithm applied for $R = 1, 2, \ldots, 349$. In most cases, fewer than seven iterations, and never more than twelve, are required to achieve convergence to eight decimal places. The convergence is slowest for R close to $R_{\dagger}$. Of course, the algorithm fails to converge for $R > R_{\dagger}$. The evaluations (about 700 applications of the Newton-Raphson algorithm with a total of about 4200 iterations) take about 0.75 s of CPU time.

In practice, the distributions for the two groups are estimated. The problem should then be solved for several sets of plausible values of the parameters of the two distributions. We can regard the interval $(c_L^{(1)}, c_L^{(2)})$ of thresholds that correspond to the limits for R as a gray zone. This gray zone can be widened by incorporating the uncertainty about the distributional parameters. Whilst we want the gray zone to be as narrow as possible, we should be honest in its construction and incorporate in it, within reason, all sources of uncertainty.

Extensions to more than two components, and to mixture distributions for the ordinary units are obvious; much of the discussion in Sect. 5.3 carries over to this setting directly. Finite mixtures can be used not only to represent subsets of the units, but simply as approximations to the distribution of the units when their distribution does not belong to any class that we can readily recognise.

Some parallels can be drawn with the search for outliers among a set of residuals, although we would be hard-pressed to define a suitable distribution for the exceptions, and would have difficulty also with specifying the frequency of the exceptional units. The purpose of the analysis, or its agenda, to seek valuable units with outstanding attributes, as opposed to searching for units that spoil the good fit of a model, plays a key role in how we proceed.

6.5 Problems, Exercises and Suggested Reading

1. Define the analogues of the functions G_0, G_1 and G_2 introduced in Sect. 6.3.1 for the normally distributed outcomes and rewrite the results for the normal outcomes in terms of these functions.

2. Compare graphically the expected losses with normally and t-distributed markers and devise some rules of thumb for the numbers of degrees of freedom k_1 and k_2 for which the normal approximation is adequate. Do the same with the optimal thresholds T.

3. Discuss the difficulties with classification using a marker with discrete distributions (e.g., Poisson) within the classes. For the Poisson, consider an approximation by a continuous distribution (find which one would fit well) and apply a method of classification in Sects. 6.1 or 6.3.
4. Rework the analysis for contaminants with t distributed markers.
5. Study (by graphics) the densities of beta distributions and devise some realistic scenarios for searching for contaminants with beta distributed markers.
6. Work out the details for classification into three categories, such as negative, positive and near-zero. Give careful consideration to the specification of the loss function(s).
7. Suggested reading: Longford (2013a,b). For the literature on screening based on the Youden index and its adaptations, see Molanes-López and Letón (2011) and references therein.
8. Look up the literature on the false discovery rate (FDR), starting with Benjamini and Hochberg (1995), and discuss how their concern carries over to the decision-theoretical setting of this chapter.
9. Search the literature for examples of classification of human subjects according to attributes such as poverty/prosperity, literacy, scholastic aptitude, homelessness, physical disability, employment, and the like. Describe the markers used and the rules based on them for classification and discuss ways of improving the classification.
10. Suppose the distribution of a variable Y is a mixture of K normals. Instead of Y we observe $Z = Y + \epsilon$, where $\epsilon \sim \mathcal{N}(0, \sigma_e^2)$ independently of Y. Find the conditional distribution of Y given Z. Relate the result to the information lost and expected loss inflated when the marker Y, with distinct normal distributions within the groups (classes), is observed subject to measurement error.
11. A project: Consider the problem of setting off a specified set of emergency measures (raising an alarm) to combat the epidemic of a disease or to deal with the threat of environmental contamination. The courses of action are to do nothing and to set off the alarm. How should the rules for the latter be set? Consider carefully the errors of the two kinds, with their effects both in the short and long term, what useful (historical) data may be available, who are the experts to be consulted, and how the elicitation should be conducted with them.
12. A project: Suppose screening for a particular disease is conducted in two stages. Some subjects are declared as definitely negative in stage I. The others attend another round of screening and only those who are found positive again are treated as cases. Simulate such a scenario on the computer. Consider carefully the distributions, loss functions and probabilities of positives and negatives in the two stages. How should the unit costs of screening at the two stages be taken into account?
13. An unsolved problem: How should the methods of this chapter be adapted to penalising, in addition to false positives and false negatives, for classifying too few or too many units as exceptional? The number of exceptional units is not known, but some prior information about it may be available.

14. Suggested reading: Literature on the analysis of microarray experiments, starting
with Speed (2003); see also Zhang and Liu (2011). Literature on fraud detection:
Bolton and Hand (2002). Literature on medical testing: Pepe (2003).

References

Benjamini, Y., & Hochberg, Y. (1995). Controlling the false discovery rate: A practical and powerful
   approach to multiple testing. *Journal of the Royal Statistical Society Series B*, *57*, 289–300.
Bolton, R., & Hand, D. (2002). Statistical fraud detection: A review. *Statistical Science*, *17*, 235–
   255.
Longford, N. T. (2013a). Screening as an application of decision theory. *Statistics in Medicine*, *32*,
   849–863.
Longford, N. T. (2013b). Searching for contaminants. *Journal of Applied Statistics*, *40*, 2041–2055.
Molanes-López, E. M., & Letón, E. (2011). Inference of the Youden index and associated thresholds
   using empirical likelihood for quantiles. *Statistics in Medicine*, *30*, 2467–2480.
Pepe, M. S. (2003). *The statistical evaluation of medical tests for classication and prediction*. New
   York: Oxford University Press.
Speed, T. (2003). *Statistical analysis of gene expression microarray data*. Boca Raton: Chapman
   and Hall/CRC.
Youden, W. J. (1950). Index for rating diagnostic tests. *Cancer*, *3*, 32–35.
Zhang, Y., & Liu, J. S. (2011). Fast and accurate approximation to significance tests in genome-wide
   association studies. *Journal of American Statistical Association*, *106*, 846–857.

Chapter 7
Small-Area Estimation

Small-area estimation (SAe) is concerned with inferences about the districts (subdomains) or another division of a country (the domain) when the subsample sizes for some subdomains are not large enough for reliable inferences about them to be based solely on the subsamples. A national survey could be designed with a sufficiently large subsample for every district of the country. However, such a survey would in some settings be prohibitively expensive. Also, SAe is sometimes conducted on data already collected in surveys that were not originally designed for that purpose.

Estimators of a district-level summary of a variable, such as the average household income, that are based only on the recorded values of the variable (household income) for the district, are called *direct*. In most applications, there is an obvious direct estimator; it is (approximately) unbiased, and among the unbiased estimators it is (nearly) efficient. Of course, the absence of bias is of dubious value when the sampling variance is large.

The key concept in SAe that facilitates improvement on direct estimation is *borrowing strength* (Efron and Morris 1972, 1973; Robbins 1955), or exploiting the similarity of the districts. The direct estimator for district d, denoted by $\hat{\theta}_d$, has a natural competitor, the national version of the estimator, denoted by $\hat{\theta}$. It is potent for a district with a small sample size, but hardly worth considering for districts with large subsamples, for which improvement on the direct estimator is of little interest.

An analyst's first instinct may be to choose one of the contending estimators, $\hat{\theta}_d$ or $\hat{\theta}$, as is the custom in model selection. This and related ideas have been abandoned in SAe a long time ago in favour of combining the estimators. A composition of estimators $\hat{\boldsymbol{\xi}} = (\hat{\xi}_1, \ldots, \hat{\xi}_K)^\top$ is defined as their linear combination $\tilde{\xi} = \mathbf{b}^\top \hat{\boldsymbol{\xi}}$, with weights in $\mathbf{b}$ that add up to unity; $\mathbf{b}^\top \mathbf{1} = 1$; $\mathbf{1}$ is the vector of unities. In a simple version of SAe, we compose $\hat{\theta}_d$ and $\hat{\theta}$ as

$$\tilde{\theta}_d = (1 - b_d)\hat{\theta}_d + b_d \hat{\theta}, \tag{7.1}$$

with the coefficients b_d, specific to district d, set so as to optimise the criterion of our choice. This criterion has almost exclusively been small mean squared error

N. T. Longford, *Statistical Decision Theory*,
SpringerBriefs in Statistics, DOI: 10.1007/978-3-642-40433-7_7,
© The Author(s) 2013

(MSE), or its variant that makes the problem easier to handle, even though there are applications which have a well recognised purpose, for which the losses are distinctly asymmetric functions of the estimation error $\tilde{\theta}_d - \theta_d$. Such an example is presented in Sect. 7.3.

The next section gives more background to SAe and outlines two established approaches. The following section discusses their common weaknesses with reference to an analysis conducted for (government) administrative decision making. We show that the established approach is poorly suited for this purpose because of the asymmetry of the losses associated with the two kinds of inappropriate decision. In the conclusion, we discuss how the practice of government statistics should be altered to respond to some new challenges.

7.1 Composition and Empirical Bayes Models

With MSE as the criterion for the estimators in (7.1), we aim to minimise the function

$$\mathrm{mse}(b_d) = \mathrm{MSE}\left\{\tilde{\theta}_d(b_d); \theta_d\right\} = (1-b_d)^2 v_d + 2b_d(1-b_d)c_d + b_d^2 v + (\theta_d - \theta)^2,$$

where $v_d = \mathrm{var}(\hat{\theta}_d)$, $c_d = \mathrm{cov}(\hat{\theta}_d, \hat{\theta})$ and $v = \mathrm{var}(\hat{\theta})$. The solution is

$$b_d^* = \frac{v_d - c_d}{v_d - 2c_d + v + (\theta_d - \theta)^2}. \tag{7.2}$$

The term $(\theta_d - \theta)^2$ is problematic when θ_d is not estimated with any appreciable precision, but that is exactly when we would like to improve on the direct estimator $\tilde{\theta}_d(0) = \hat{\theta}_d$ most. The problem is resolved by replacing $(\theta_d - \theta)^2$ with its average over the districts, the district-level variance

$$\sigma_{\mathrm{B}}^2 = \mathrm{var}_d(\theta_d) = \frac{1}{D}\sum_{d=1}^{D}(\theta_d - \theta)^2. \tag{7.3}$$

It is estimated by moment matching; see Appendix. When the sample size for district d is much smaller than the overall (national) sample size n, then $v_d \gg v$ and $v_d \gg c_d$. In that case, the error caused by dropping v and c_d from the expression for b_d^* in (7.2) is negligible. Hence the approximation

$$b_d^* \doteq \frac{v_d}{v_d + \sigma_{\mathrm{B}}^2}, \tag{7.4}$$

in which the replacement of $(\theta_d - \theta)^2$ by σ_{B}^2 should be the main concern, except for one or two districts for which n_d is a substantial fraction of n, for which v and c_d cannot be ignored.

We obtain the fraction in (7.4), seemingly without any approximation, by appealing to the model

$$y_{id} = \theta + \delta_d + \varepsilon_{id} \qquad (7.5)$$

for (normally distributed) outcomes y for units i in districts d, where δ_d and ε_{id} are two independent random samples from centred normal distributions with respective variances σ_B^2 and σ_W^2. Suppose θ and the variances σ_B^2 and σ_W^2 are known. Then the conditional distribution of $\delta_d = \theta_d - \theta$ given the outcomes $\mathbf{y}_d = (y_{1d}, \ldots, y_{n_d d})^\top$ is

$$\mathcal{N}\left(\frac{\sigma_B^2}{\sigma_W^2 + n_d \sigma_B^2}(\mathbf{y}_d - \theta \mathbf{1}_{n_d})^\top \mathbf{1}_{n_d}, \ \frac{\sigma_B^2 \sigma_W^2}{\sigma_W^2 + n_d \sigma_B^2}\right). \qquad (7.6)$$

The conditional expectation coincides with the approximation $(1 - b_d^*)\hat{\theta}_d + b_d^*\theta$ using (7.4) when $\hat{\theta}_d$ is the sample mean for district d. For data with distribution other than normal, generalised mixed linear models (McCullagh and Nelder 1989) have to be applied. They entail some computational complexity.

On the one hand, we have the comfort of a respectable model in (7.5); on the other, we have with (7.1) the freedom from distributional assumptions and no constraints on how the estimator $\hat{\theta}_d$ is defined, so long as it is unbiased and we know (or can reliably estimate) its sampling variance v_d. The estimator derived from the model in (7.5) has some commonality with Bayes methods. If σ_B^2 were known and regarded as the prior variance of the D (unknown) quantities θ_d, the estimator $\tilde{\theta}_d$ would be the posterior expectation of θ_d. As σ_B^2 is estimated, we refer to $\tilde{\theta}_d$ obtained from (7.6) as an *empirical Bayes* estimator. The estimator derived through (7.2) or (7.4) is a composite estimator. Thus far they coincide, although they are associated with different assumptions. The composition can be interpreted as a *shrinkage*, pulling the unbiased (but incompletely informed) estimator $\hat{\theta}_d$ toward the biased but stable alternative $\hat{\theta}$. More shrinkage is applied when $\hat{\theta}_d$ is estimated with less precision.

It is obvious that $\tilde{\theta}_d$ is particularly effective when σ_B^2 is small in relation to σ_W^2, when the districts are similar. In the extreme case, when $\sigma_B^2 = 0$, θ_d is estimated with very high precision by $\tilde{\theta}_d = \hat{\theta}$ for all districts d. In contrast, when σ_B^2 is very large, $b_d^* \doteq 0$ for all d and $\tilde{\theta}_d$ differs from $\hat{\theta}_d$ only slightly, because $\hat{\theta}$ may be a very poor estimator of θ_d.

The differences among the districts may be greatly reduced, and their similarity enhanced, by adjustment for suitable covariates. That is, the variance $\sigma_B^2 = \mathrm{var}_d(\delta_d)$ may be much smaller in a regression model

$$\mathbf{y}_d = \mathbf{X}_d \boldsymbol{\beta} + \delta_d \mathbf{1}_{n_d} + \boldsymbol{\varepsilon}_d, \qquad (7.7)$$

with suitable covariates $\mathbf{X}$. This advantage over (7.5) is matched by the multivariate version of the composition. Let $\boldsymbol{\theta}_d$ be a $K \times 1$ vector of some population quantities related to district d, and suppose θ_d, the quantity of interest, is its first element. So, $\theta_d = \mathbf{u}^\top \boldsymbol{\theta}_d$, where $\mathbf{u} = (1, 0, \ldots, 0)^\top$ identifies the first element. Let $\boldsymbol{\theta}$ be the national counterpart of θ_d. Further, let $\hat{\theta}_d$ be a vector of direct estimators of the elements in θ_d and $\hat{\theta}$ an unbiased estimator of θ. The multivariate composition of $\tilde{\theta}_d$ is defined as

$$\tilde{\theta}_d = (\mathbf{u} - \mathbf{b}_d)^{\top} \hat{\theta}_d + \mathbf{b}_d^{\top} \hat{\theta},$$

with suitable vectors $\mathbf{b}_d$. In fact, they are set so as to minimise the MSE of the (multivariate) composite estimator $\tilde{\theta}_d$. The solution is the multivariate analogue of (7.2), $\mathbf{b}_d^* = \mathbf{Q}_d^{-1} \mathbf{P}_d \mathbf{u}$, where

$$\begin{aligned} \mathbf{Q}_d &= \mathbf{V}_d - \mathbf{C}_d - \mathbf{C}_d^{\top} + \mathbf{V} + \mathbf{B}_d \mathbf{B}_d^{\top} \\ \mathbf{P}_d &= \mathbf{V}_d - \mathbf{C}_d \end{aligned} \tag{7.8}$$

and $\mathbf{B}_d = \theta_d - \theta$, $\mathbf{V}_d = \mathrm{var}(\hat{\theta}_d)$, $\mathbf{C}_d = \mathrm{cov}(\hat{\theta}_d, \hat{\theta}_d)$ and $\mathbf{V} = \mathrm{var}(\hat{\theta})$.

All the derivations, for both empirical Bayes and composite estimators are wedded to the criterion of minimum MSE. There is only one deviation, which corresponds to the averaging that introduces σ_{B}^2 by (7.3), or its multivariate version $\boldsymbol{\Sigma}_{\mathrm{B}}$, for composite estimation. By considering the averaged MSE, it resolves the conflict between the assumption of randomness of δ_d in the model and the sampling-design view that θ_d is fixed. The random terms δ_d are meant to vary across replications. However, in our context they correspond to well identified (labelled) districts that have patently fixed targets θ_d. As a consequence, the standard error derived from the variance in (7.6), or its version for (7.7), is correct only for districts with $\delta_d = \pm \sigma_{\mathrm{B}}$. If $|\delta_d| > \sigma_{\mathrm{B}}$, the estimate of the standard error based on (7.6) is too small (optimistic); if $|\delta_d| < \sigma_{\mathrm{B}}$, it is too pessimistic. To see this, consider an 'average' district, for which $\theta_d \doteq \theta$. For it, $\tilde{\theta}_d$ is a composition of two unbiased estimators, so it is also unbiased. For a district with an exceptional value of θ_d, but the same variance v_d, we obtain the composition with the same coefficients, and therefore the same variance, but it is burdened also with the bias B_d. For details, and a proposal that addresses this problem, see Longford (2007). For more background to SAe, refer to Rao (2003) and Longford (2005).

7.2 Estimation for a Policy

Concerned about illiteracy, the Ministry of Education in a developing country allocates some funds for a programme to combat it. Illiteracy is endemic in the country, and its eradication is not a realistic goal for the time being. The country, comprising $D = 72$ districts, has 38 million adults for whom the binary variable indicating literacy is well defined by a simple questionnaire. The Ministry decides to fund a programme in every district d in which the rate of illiteracy exceeds 25 %. A district with an illiteracy rate of $\theta_d > T = 0.25$ and population size (number of adults) N_d is planned to receive a grant of $U N_d (\theta_d - T)$, where U is a monetary value, the unit cost of the programme pro-rated for an illiterate individual in excess of the critical rate T. The national rate of illiteracy is estimated as 16.15%, so the programme is aimed at a minority of the districts or, more precisely, at the districts with overall population much smaller than half of the country. The illiteracy rate in each district will be estimated from a national survey with sample size 17,500. The survey has a stratified simple random sampling design, with stratification at the district level and within-district sample sizes approximately proportional to the district's population sizes N_d.

The estimates $\hat{\theta}_d$ based on the survey lead to incorrect action when $\hat{\theta}_d > T > \theta_d$ and when $\hat{\theta}_d < T < \theta_d$, referred to as false positive and false negative cases, respectively. The illiteracy rates θ_d may be estimated with much greater precision later, so the Ministry may be held to account for the erroneous assessments based on $\hat{\theta}_d$. For illustration, we consider the piecewise absolute, linear and quadratic loss functions defined for each district d as follows:

1. $L_+(\hat{\theta}_d, \theta_d) = 1$ when $\hat{\theta}_d > T > \theta_d$ and $L_-(\hat{\theta}_d, \theta_d) = R$ when $\hat{\theta}_d < T < \theta_d$;
2. $L_+(\hat{\theta}_d, \theta_d) = \hat{\theta}_d - \theta_d$ when $\hat{\theta}_d > T > \theta_d$ and $L_-(\hat{\theta}_d, \theta_d) = R(\theta_d - \hat{\theta}_d)$ when $\hat{\theta}_d < T < \theta_d$;
3. $L_+(\hat{\theta}_d, \theta_d) = (\hat{\theta}_d - \theta_d)^2$ when $\hat{\theta}_d > T > \theta_d$ and $L_-(\hat{\theta}_d, \theta_d) = R(\hat{\theta}_d - \theta_d)^2$ when $\hat{\theta}_d < T < \theta_d$;

in all other cases, $L_+(\hat{\theta}_d, \theta_d) = 0$ or $L_-(\hat{\theta}_d, \theta_d) = 0$. Thus, the loss function is

$$L(\hat{\theta}_d, \theta_d) \;=\; I_{\hat{\theta}_d > T > \theta_d} L_+(\hat{\theta}_d, \theta_d) \;+\; R\, I_{\hat{\theta}_d < T < \theta_d} L_-(\hat{\theta}_d, \theta_d),$$

where I is the indicator function; $I_a = 1$ if the statement a is logically true, and $I_a = 0$ otherwise. From the context, it is obvious that $R \gg 1$.

Suppose we have an estimator $\hat{\theta}_d \sim \mathcal{N}(\gamma_d, v_d^2)$; $\hat{\theta}_d$ may be biased for θ_d ($\gamma_d \neq \theta_d$). Let $\tilde{z}_d = (\gamma_d - T)/v_d$ and $\tilde{z}_d^{\dagger} = (\gamma_d - \theta_d)/v_d$. For the piecewise constant loss function, we have the expected losses

$$Q_+ = \frac{1}{v_d} \int_T^{+\infty} \phi\left(\frac{y - \gamma_d}{v_d}\right) dy \;=\; \Phi(\tilde{z}_d)$$

$$Q_- = \frac{R}{v_d} \int_{-\infty}^{T} \phi\left(\frac{y - \gamma_d}{v_d}\right) dy \;=\; R\{1 - \Phi(\tilde{z}_d)\}$$

when $\theta_d < T$ and $\theta_d > T$, respectively. For the piecewise linear loss function, we have

$$Q_+ = \frac{1}{v_d} \int_T^{+\infty} (y - \theta_d) \phi\left(\frac{y - \gamma_d}{v_d}\right) dy \;=\; v_d\left\{ z_d^{\dagger} \Phi(\tilde{z}_d) + \phi(\tilde{z}_d)\right\}$$

$$Q_- = \frac{R}{v_d} \int_{-\infty}^{T} (\theta_d - y) \phi\left(\frac{y - \gamma_d}{v_d}\right) dy \;=\; R v_d\left[-z_d^{\dagger}\{1 - \Phi(\tilde{z}_d)\} + \phi(\tilde{z}_d)\right]$$

when $\theta_d < T$ and $\theta_d > T$, respectively. For the piecewise quadratic loss function,

$$Q_+ = \nu_d^2 \left\{ \left(1 + z_d^{\dagger 2}\right) \Phi(\tilde{z}_d) + \left(2z_d^\dagger - \tilde{z}_d\right) \phi(\tilde{z}_d) \right\}$$

$$Q_- = R\nu_d^2 \left[\left(1 + z_d^{\dagger 2}\right) \{1 - \Phi(\tilde{z}_d)\} - \left(2z_d^\dagger - \tilde{z}_d\right) \phi(\tilde{z}_d) \right]$$

when $\theta_d < T$ and $\theta_d > T$, respectively. The expected losses Q_+ and Q_- involve the targets θ_d through $\tilde{z}_d$, so finding which of them is smaller is difficult. Instead, we apply a scheme that is derived intuitively and has no solid theoretical foundation. However, we show by simulations that it is far superior to the established estimators which aim to minimise the MSE.

We seek the estimator that satisfies the following two conditions:

1. equilibrium at T—for a district with rate $\theta_d = T$, the choice between the two courses of action should be immaterial in expectation. That is,

$$\mathrm{E}\left\{L_+\left(\hat{\theta}_d, T\right)\right\} = \mathrm{E}\left\{L_-\left(\hat{\theta}_d, T\right)\right\};$$

2. minimum averaged MSE,

$$\min \mathrm{E}_d \left\{ \mathrm{MSE}\left(\hat{\theta}_d; \theta_d\right) \right\},$$

where the expectation (averaging) is taken over the D quantities θ_d. In the model-based approach, it corresponds to averaging over the distribution of δ_d or δ_d.

Under the equilibrium condition, $\tilde{z}_d$ and $z_d^\dagger$ coincide and the balance equation $Q_+ = Q_-$ reduces for the respective piecewise constant, linear and quadratic losses to equations familiar from the earlier chapters:

$$\Phi(\tilde{z}_d) = \frac{R}{R+1}$$

$$(R-1)\{\tilde{z}_d \Phi(\tilde{z}_d) + \phi(\tilde{z}_d)\} - R\tilde{z}_d = 0$$

$$(R+1)\left\{\left(1 + \tilde{z}_d^2\right)\Phi(\tilde{z}_d) + \tilde{z}_d \phi(\tilde{z}_d)\right\} - R\left(1 + \tilde{z}_d^2\right) = 0.$$

We solve the appropriate equation for each district d, obtaining the equilibrium value $\tilde{z}_d^*$, and then search in the class of compositions

$$\tilde{\theta}_d = (1 - b_d)\hat{\theta}_d + b_d F_d$$

for estimators that satisfy the two conditions. Here $\hat{\theta}_d$ is the direct estimator.

Assuming that $\hat{\theta}_d$ is unbiased for θ_d,

$$\mathrm{MSE}\left(\tilde{\theta}_d; \theta_d\right) = (1 - b_d)^2 v_d + b_d (F_d - \theta_d)^2. \tag{7.9}$$

To avoid any dependence on the target θ_d, we replace the term $(F_d - \theta_d)^2$ by its average over the districts, $\sigma_B^2 + (F_d - \theta)^2$, to obtain the averaged MSE,

$$\mathrm{aMSE}\left(\tilde{\theta}_d \,;\, \theta_d\right) = (1 - b_d)^2 v_d + b_d \left\{\sigma_B^2 + (F_d - \theta)^2\right\},$$

where $\sigma_B^2 = E_d\{(\theta_d - \theta)^2\}$. This variance also has to be estimated, but this can be done with reasonable precision when there are many districts and a substantial fraction of them have non-trivial subsample sizes.

The equilibrium condition implies that for a district with $\theta_d = T$,

$$\tilde{z}^* = \frac{\gamma_d - \theta_d}{v_d} = \frac{b_d(F_d - T)}{|1 - b_d|\sqrt{v_d}},$$

and hence

$$F_d = T + \frac{|1 - b_d|}{b_d}\,\tilde{z}^*\,\sqrt{v_d}. \tag{7.10}$$

With this constraint,

$$\tilde{\theta}_d = (1 - b_d)\,\hat{\theta}_d + b_d\,T + \tilde{z}^*|1 - b_d|\sqrt{v_d}$$

and

$$\mathrm{aMSE}\left(\tilde{\theta}_d \,;\, \theta_d\right) = (1 - b_d)^2 \left(1 + \tilde{z}_d^{*2}\right) v_d + b_d^2 \left\{\sigma_B^2 + (T - \theta)^2\right\}$$
$$+ 2b_d|1 - b_d^*|(T - \theta)\tilde{z}^*\sqrt{v_d}.$$

This function of b_d attains its minimum or maximum when

$$b_d^* = \frac{v_d\left(1 + \tilde{z}_d^{*2}\right) - \mathrm{sign}(1 - b_d^*)(T - \theta)\tilde{z}_d^*\sqrt{v_d}}{v_d + \sigma_B^2 + \{\tilde{z}_d^*\sqrt{v_d} - \mathrm{sign}(1 - b_d^*)(T - \theta)\}^2}\,;$$

the sign function is equal to $+1$ for positive arguments, -1 for negatives and $\mathrm{sign}(0) = 0$. The aMSE has an odd number of extremes because it diverges to $+\infty$ when $b_d \to \pm\infty$. But the form of b_d^* indicates that it can have at most two extremes; therefore its minimum is unique. Without the equilibrium condition, the MSE in (7.9) attains its minimum in the range $(0,1)$. With the condition, this is no longer the case. For example, b_d^* is negative when

$$\sqrt{v_d} < \frac{z_d^*}{1 + z_d^{*2}}(T - \theta).$$

Neither is b_d^* increasing in v_d, anticipating that more information about the district would be associated with less shrinkage. No shrinkage is applied when θ_d is known, when $v_d = 0$, but also when $\sqrt{v_d} = (T - \theta)\tilde{z}_d^*/(1 + \tilde{z}_d^{*2})$.

When $b_d^* = 0$, F_d in (7.10) is not defined. However, the product $b_d^* F_d$, and therefore the estimator $\tilde{\theta}_d$, are well defined by its limit as $b_d^* \to 0$, equal to $\tilde{z}_d^*\sqrt{v_d}$ and $\hat{\theta}_d + \tilde{z}_d^*\sqrt{v_d}$, respectively.

7.3 Application

In this section, we describe the assessment of the estimators $\tilde{\theta}_d$ by a simulation study. We construct an artificial country, with its division into districts, and replicate the processes of sampling and estimation, using the alternative estimators $\hat{\theta}_d$ (direct), $\tilde{\theta}_d$ (minimum aMSE) and $\tilde{\theta}_d^*$, based on the linear loss with a range of penalty ratios R. The population sizes of the districts are N_d, and their total, the national population size is N.

We estimate D quantities, one per district, and so we have to be prepared for the conclusion that one estimator is better for some districts and another better for some others. However, the loss function we adopt offers a simple comparison by the (weighted) total of the district-level expected losses.

The studied country has an adult population of 37.85 million. The districts vary a great deal in their population sizes. The least populous districts have just over 50,000 adults, and three (city) districts have over 1.5 million adults each. Two other districts have population of adults in excess of 1.5 million each. The rates of illiteracy tend to be lower in the most populous districts, but they are low also in some of the least populous districts, in which a large percentage of the population lives in a single mid-size town.

The (adult) population sizes and rates of illiteracy in the districts are displayed in Fig. 7.1. The sizes of the discs are proportional to the district's sample sizes. Note the relatively large difference between the national rate $\theta = (N_1\theta_1 + \cdots + N_D\theta_D)/N = 16.15\%$ and the mean of the district-level rates, $\bar{\theta} = (\theta_1 + \cdots + \theta_D)/D = 18.0\%$.

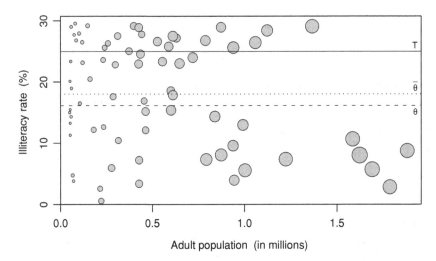

Fig. 7.1 The rates of illiteracy in the districts of the studied country. The size of the disc is proportional to the subsample size of the district in a national survey with a stratified sampling design. The horizontal lines mark the threshold, $T = 25.0\%$, the average of the district-level rates, $\bar{\theta} = 18.0\%$, and the national rate, $\theta = 16.15\%$

The rates of illiteracy are in the range $0.6-29.6\%$. Twenty-four districts have rates above the threshold $T = 25\%$; their total population is 11.63 million (30.7%). We refer to them as the *deserving* districts.

We generated the data on which Fig. 7.1 is based by a random process, informed by the state of affairs in a particular country. In practice, the population sizes may be known, or estimated with reasonable precision, but the illiteracy rates would not be; otherwise there would be no rationale for a survey. The district-level sample sizes are in the range 22–870. They are set not exactly proportionally to the population sizes, to reflect other aims of the survey, nonresponse, the resources available, and various difficulties and constraints in the conduct of the survey.

We assume the piecewise linear loss function with plausible penalty ratios in the range (15, 40); implementing the policy where it is not intended is regarded as a much less serious error than failing to implement it where it should be. At present, we are not concerned about the waste of Ministry's resources on districts that are not deserving, because these expenditures are not considered as yet. The consequences considered are the Ministry's loss of face and credibility when the rates of illiteracy are estimated much more precisely, or revealed in the future, and the failure of the programme to deal with endemic illiteracy. Section 7.3.1 addresses the issue of limited resources.

Table 7.1 summarises the results of a single replication of the processes of sampling and estimation, using the linear loss functions with $R = 15$ and $R = 40$. The pairs of columns give the numbers of districts and their total population with errors of the two kinds for the four estimators. The first two pairs are in no way flattering the estimators based on (linear) loss functions — these estimators involve more districts, 14 for $R = 15$ and 16 for $R = 40$, and only slightly less population, 3.57 and 3.73 million, than the direct and composite estimators (11 and 12 districts, with 3.62 and 3.80 million, respectively). However, by the criterion that we regard as principal, the linear loss with $R = 15$ or $R = 40$, the direct and composite estimators perform very poorly. The estimators based on the linear loss have losses 0.34 and 0.54 for $R = 15$ and $R = 40$, respectively, whereas the corresponding losses with the direct

Table 7.1 The losses in a replication of sampling and estimation for the linear loss function with $R = 15$ and $R = 40$, summarised separately for the deserving (false negative, F−) and not deserving (false positive, F+) districts. For F−, separate values of the loss are given for $R = 15$ and $R = 40$

Estimator (R)	Districts		Population		Loss	
	F−	F+	F−	F+	F−	F+
MinLoss (15)	2	12	0.183	3.389	0.168	0.170
MinLoss (40)	2	14	0.183	3.548	0.328	0.213
Direct $\binom{15}{40}$	5	6	1.767	1.859	1.141 3.042	0.055
Composite $\binom{15}{40}$	8	4	2.232	1.573	1.412 3.765	0.028

Table 7.2 The average losses in 1,000 replications of sampling and estimation for the linear loss function with $R = 15, 25$ and 40, for the deserving (F–) and not deserving (F+) districts. The averages are for the numbers of districts (Districts), the adult population of these districts (Population), and the losses (Loss)

Estimator (R)	Districts		Population		Loss	
	F–	F+	F–	F+	F–	F+
MinLoss (15)	1.59	10.32	0.525	2.793	0.233	0.146
MinLoss (25)	1.16	11.38	0.382	3.084	0.271	0.171
MinLoss (40)	0.96	12.01	0.284	3.304	0.309	0.191
Direct (15)	6.43	3.75	2.401	1.068	1.333	0.043
Composite (15)	9.59	2.13	3.239	0.768	1.828	0.023

estimator are 1.20 and 3.10 and with the composite estimator 1.44 and 3.79. For the direct and composite estimators, the losses for false negatives (F–) are in proportion 15 : 40 for the respective penalty ratios $R = 15$ and 40, whereas the losses with false positives are not changed.

The direct and composite estimators incur most of their losses in the deserving districts (false negatives), and these losses are severe owing to the high penalty ratio. In contrast, the decision-theory based estimators reduce their exposure to such errors to minimum, at the price of more errors on false positives. However, the latter errors are much less costly.

Table 7.2 presents the averages over 1,000 replications for the summaries listed in Table 7.1. For the direct and composite estimators, the expected loss is listed only for $R = 15$; the respective values for $R = 25$ and 40 are obtained by multiplying it by 1.67 and 2.67. The table confirms our conclusion from a single replication. While we may argue over the details of the loss function, or its plausible range, it is difficult to defend the established concensus that the composite estimator is suitable for all purposes. In fact, in our simulations it is inferior even to the direct estimator, by a nontrivial margin.

The advantage of the estimators tailored to the purpose is so great that we can afford a wide range of plausible penalty ratios R, and still incur much smaller expected losses than with the direct or composite estimators. Figure 7.2 displays the expected losses and the frequencies of incorrect decisions for the districts. Each district is associated with five vertical bars. Three of them, each topped by a black disc, correspond to the decisions based on the linear loss with $R = 15, 25$, and 40, from left to right. Two other bars are for the decisions based on the direct estimator (left) and the composite estimator (right). The horizontal ticks on these bars mark the expected loss with $R = 15$ (lowest), 25 and 40 (at the top).

The top panel shows that the losses due to false positive assessments (in 48 districts) are very small compared to the losses due to false negatives (24 districts). Simply, the direct and composite estimators are caught out by the specific purpose intended for their use. The bottom panel shows that the decisions based on minimum expected loss are not particularly effective for another criterion. If we merely count

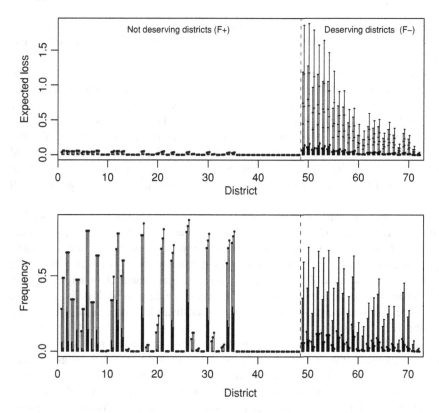

Fig. 7.2 The empirical expected losses and the frequencies of incorrect decisions for the districts. The districts are in the ascending order of (adult) population size within the groups of not deserving and deserving districts, separated by the vertical dashes. The horizontal ticks mark the expected losses for the direct (*left*) and composite estimators (*right*) at $R = 15$, 25 and 40. The expected losses for the decision-theory based estimators are marked by black discs at the top of the segment, ordered by R from left to right

the errors committed in the decisions, there is a clear division. Many false positives are generated, especially for the least populous districts, but the consequences of these errors are negligible when compared to the false negatives. There the direct and composite estimators are in a distinct disadvantage that is decisive for the overall comparison.

7.3.1 Limited Budget

Thus far, we were concerned solely with the consequences of incorrect decisions in terms of misdirecting the programme to non-deserving districts, and failing to apply it in deserving districts. Another important factor is the programme's budget. It may be insufficient even if all the deserving districts were correctly identified. When the classification of the districts is subject to uncertainty, the minimum-expected-loss solution appears to be generous, if not profligate, preferring to err on the side of false positives.

Denote by A_d the award that district d should receive. It is $A_d = 0$ for non-deserving districts and $A_d = U N_d(\theta_d - T)$ for deserving districts. If these quantities are known and $A_1 + \cdots + A_D$ exceeds the Ministry's budget, F, then at least some of the deserving districts have to be short-changed. The award could be cut by the same percentage for every district, the threshold T could be raised, the award could be withheld completely from some of the (deserving) districts, or the shortfall could be shared equally by all the deserving districts. Denote by G_d the actual award to districts d. The shortfall is $S_d = A_d - G_d$; we assume that it is nonnegative for all districts.

We define a penalty function, such as $\sum_d (A_d - G_d)^2$, and minimise it subject to the constraint that $G_1 + \cdots + G_d \leq F$. Such an approach can be applied to the estimated awards without any changes, although we could not refer to any optimality properties of the resulting allocation, because they may not carry over from the setting with certainty, when all θ_d are known. In any case, we cannot claim any optimality properties for the allocation based on our procedure because it does not minimise the expected loss, merely performs much better than the allocation based on some established alternatives.

If the Ministry has a single budget from which both the survey and the programme are to be funded, then it does not have to commit itself to a given design, but can decide how much to spend on the survey. Higher spending on the survey results in greater sample sizes and more information about θ_d for every district, although this is accompanied by reduced funds for the awards to the selected districts. There is no analytical solution to this problem, but the alternative scenarios are easy to simulate. See Longford (2011, 2012b) for examples discussed in detail. One important input into such a simulation entails relating the losses due to false positives and negatives to the losses due to imperfect implementation of the programme (under-funding). This is a difficult accounting exercise. It can be foregone by presenting the client with a range of options (sample sizes of the surveys and expected losses generated on a mock-up of the country).

7.4 Conclusion

The study presented in this chapter demonstrates that separating the tasks of (statistical) estimation and (administrative) implementation based on the estimates is a bad strategy. When the estimation is competent (e.g., efficient, or at least nearly so), and the administration of the programme would be competent if the uncertainty about the estimates could be ignored, the combination of the two processes is suboptimal. We propose to integrate the two processes and address the distribution of the funds to the districts as the ultimate statistical problem. In principle, if two clients are interested in the same set of inferential targets (population quantities), but intend to use them for different purposes, or with different perspectives, we should not hesitate to give them two different answers. Consistency, giving the same answers to both clients, is a superficial quality that is unlikely to serve the interests of both of them.

The separation of the tasks of (government) policy formulation and implementation on the one hand and survey design and analysis on the other is a result of the desire to have independent statistical advice informing policy makers, that is, noninterference and objectivity. The development in this chapter indicates that that leads to a suboptimal practice, because the analysis has to be informed by the purpose for which the estimates will used.

7.5 Appendix: Estimating σ_B^2

Suppose we have district-level direct (unbiased) estimators $\hat{\theta}_d$ with respective variances v_d; their national counterparts are $\hat{\theta}$ and v. The covariance of $\hat{\theta}_d$ and $\hat{\theta}$ is denoted by c_d. We form the statistic

$$S_B = \frac{1}{D} \sum_{d=1}^{D} \left(\hat{\theta}_d - \hat{\theta} \right)^2,$$

derive its expectation, and define an estimator of σ_B^2 by matching the expectation of S_B with its realised value. The expectation of S_B is

$$E(S_B) = \frac{1}{D} \sum_{d=1}^{D} \left\{ v_d - 2c_d + v + (\theta_d - \theta)^2 \right\}$$

$$= \frac{1}{D} \sum_{d=1}^{D} (v_d - 2c_d) + v + \sigma_B^2. \qquad (7.11)$$

Hence the estimator

$$\hat{\sigma}_B^2 = S_B - \frac{1}{D} \sum_{d=1}^{D} (v_d - 2c_d) - v.$$

It can be interpreted as the naive estimator S_B adjusted for the uncertainty about θ_d and θ. When $\hat{\theta} = (w_1 \hat{\theta}_1 + \cdots + w_D \hat{\theta}_D)/w_+$ for suitable (positive) constants w_d and $w_+ = w_1 + \cdots + w_D$, as in standard estimators of the population mean or total, then $c_d = w_d v_d / w_+$.

In (7.11) we assumed that θ is the mean of θ_d. However, θ is their *population-weighted* mean. The 'unweighted' mean $\bar{\theta}$ differs from it substantially in our example, by nearly 2%. In practice, θ is used because it is estimated with greater precision, as the influence of least populous districts, whose $\hat{\theta}_d$ is estimated with least precision, is reduced. We use $\bar{\bar{\theta}}$ throughout.

7.6 Problems, Exercises and Suggested Reading

1. Derive the expressions in (7.2) and (7.8).
2. Study the properties of the estimator in (7.2) when we use a coefficient b_d other than the optimal. How much less efficient are $\tilde{\theta}_d(b_d^* + a)$ and $\tilde{\theta}_d(b_d^* - a)$ than $\tilde{\theta}_d(b_d^*)$ for a small constant a? What strategy in estimating b_d^* does this suggest?
3. Derive from Tables 7.1 and 7.2 the (average) losses with misspecified penalty ratio R. That is, suppose a particular value R was used in making the relevant decisions, but you would like to assess the result with a different penalty ratio R'. In particular, if we use the decisions based on $R = 40$, but evaluate them with $R' = 15$, do we still incur much lower losses in expectation than with the direct and composite estimators?
4. Summarise the advice regarding the conduct of the analysis of variance that is implied by small-area estimation. As an example, consider the setting with a small and a large group (and some other groups), and discuss estimation of their expectations.
5. Discuss the difficulties that would be encountered in the direct minimisation of the expected loss for a small area.
6. Suggested reading: Fay and Herriot (1979)—a seminal paper that marks the beginning of modern SAe; Ghosh and Rao (1994)—a review of model-based SAe; Shen and Louis (1998)—a discussion of purpose-specific estimation for small areas; Longford (2004) and Molina et al. (2007)—estimation of unemployment rates in the UK Labour Force Survey; Longford (2012a)—the issue of limited budget.
7. In small-area estimation we combine alternative estimators. Why not do the same instead of model selection? See Longford (2005, 2012b).
8. A project: Implement the estimators discussed in Shen and Louis (1998) and reproduce by simulations their conclusion that different estimators are superior for different inferential goals (θ_d, its ranks, and the distribution of the values of θ_d). Other relevant goals may be considered, the dispersion of θ_d and their extremes in particular.
9. A project: Find an application of small-area estimation in your or a neighbouring country, together with the purpose for which it is used. Define plausible ranges of the losses emanating from the estimation errors (or inappropriate decisions) and address the problem of minimising their total in expectation first informally, and then by the method of Sect. 7.2.
10. Further applications of small-area estimation: disease mapping (Lawson 2008); social sciences (Congdon 2010); data in space and time (Longford 2010; Cressie and Wikle 2011).

References

Congdon, P. (2010). *Applied Bayesian hierarchical methods*. London: Chapman and Hall/CRC.

Cressie, N., & Wikle, C. K. (2011). *Statistics for spatio-temporal data*. New York: Wiley.

Efron, B., & Morris, C. N. (1972). Limiting the risk of Bayes and empirical Bayes estimators. Part II: The empirical Bayes case. *Journal of the American Statistical Association, 67*, 130–139.

Efron, B., & Morris, C. N. (1973). Stein's estimation rule and its competitors: An empirical Bayes approach. *Journal of the American Statistical Association, 68*, 117–130.

Fay, R. E., & Herriot, R. A. (1979). Estimates of income for small places: An application of the James-Stein procedures to census data. *Journal of the American Statistical Association, 74*, 269–277.

Ghosh, M., & Rao, J. N. K. (1994). Small area estimation. An appraisal. *Statistical Science, 9*, 55–93.

Lawson, A. B. (2008). *Bayesian disease mapping: Hierarchical modelling in spatial epidemiology*. Boca Raton: CRC Press.

Longford, N. T. (2004). Missing data and small area estimation in the UK labour force survey. *Journal of the Royal Statistical Society Series A, 167*, 341–373.

Longford, N. T. (2005). *Missing data and small-area estimation. Analytical equipment for the survey statistician*. New York: Springer.

Longford, N. T. (2007). On standard errors of model-based small-area estimators. *Survey Methodology, 33*, 69–79.

Longford, N. T. (2010). Small-area estimation with spatial similarity. *Computational Statistics and Data Analysis, 54*, 1151–1166.

Longford, N. T. (2011). *Policy-related small-area estimation*. CEPS Working Paper No. 2011–44. CEPS/INSTEAD, Esch-sur-Alzette, Luxembourg, 2011. Retrieved from http://www.ceps.lu/pdf/3/art1662.pdf

Longford, N. T. (2012a). Allocating a limited budget to small areas. *Journal of Indian Society for Agricultural Statistics, 66*, 31–41.

Longford, N. T. (2012b). Which model..? is the wrong question. *Statistica Neerlandica, 66*, 237–252.

McCullagh, P., & Nelder, J. A. (1989). *Generalized linear models* (2nd ed.). London: Chapman and Hall.

Molina, I., Saei, A., & Lombardía, M. J. (2007). Small area estimates of labour force participation under a multinomial logit mixed model. *Journal of the Royal Statistical Society Series A, 170*, 975–1000.

Rao, J. N. K. (2003). *Small area estimation*. New York: Wiley.

Robbins, H. (1955). An empirical Bayes approach to statistics. In J. Neyman (Ed.), *Proceedings of the Third Berkeley Symposium on Mathematical Statistics and Probability 1* (pp. 157–164). Berkeley, CA: University of California Press.

Shen, W., & Louis, T. A. (1998). Triple-goal estimates in two-stage hierarchical models. *Journal of the Royal Statistical Society Series B, 60*, 455–471.

Chapter 8
Study Design

The previous chapters were almost exclusively concerned with data analysis. We frequently analyse datasets that were collected without our input as to how the observational units or experimental subjects should be selected, how many of them, which measurement instruments and protocols should be applied, how the recorded information should be coded, and other details. Many analyses are secondary, related to issues other than the original (and the principal) motive for the study.

The subject of this chapter is design—the plan for data collection that would ensure or promote the goal of drawing inferences of prescribed quality with the available resources. It entails a balancing act between collecting sufficient information and being frugal with all the resources involved (funding, time, involvement of human subjects and environment in general).

To proceed systematically, we have to establish the cost of the study as a function of the factors that can be manipulated in the design (sample size in particular), and the value of the inference as a function of the sampling variation (the cost of uncertainty), expected loss, probability of an unequivocal conclusion, and the like. As an alternative, we can present the client who plans to conduct the study with these two functions, and let him choose the design settings, without formulating the criteria for selection. Everybody with good reason would like to have least uncertainty at the lowest cost of experimentation. We regard the cost of analysis, as well as the cost of design (planning), as negligible—analysts' time, effort and their equipment are usually much cheaper than the experiments they are designing.

We focus on the simplest problem of comparing two samples, on which the principles can be elaborated with least distraction. For other problems, a number of limitations have to be reckoned with, foremost the need to reconcile several purposes of the study and external constraints (imperatives) that are specific to the study and cannot be treated by a general approach. First we review the established approach based on hypothesis testing.

N. T. Longford, *Statistical Decision Theory*,
SpringerBriefs in Statistics, DOI: 10.1007/978-3-642-40433-7_8,
© The Author(s) 2013

8.1 Sample Size Calculation for Hypothesis Testing

Suppose we plan to compare two normally distributed samples, both with known variance σ^2, and intend to test the hypothesis that the two expectations, μ_1 and μ_2, are identical. We set the size of the test to α (usually to 0.05), and would like to achieve power β (usually 0.80 or 0.90) when the difference $\Delta\mu = \mu_2 - \mu_1$ is equal to a given value Δ. We consider the one-sided hypothesis, with the alternative formed by the positive values of $\Delta\mu$.

Suppose the within-group sample sizes n_1 and n_2 have to be such that $n_2 = rn_1$. Denote $m = 1/n_1 + 1/n_2$. The test statistic $z = (\hat{\mu}_2 - \hat{\mu}_1)/\sqrt{m}/\sigma$ has the distribution $\mathcal{N}(\Delta\mu/\sqrt{m}/\sigma, 1)$. The critical region is the interval $(c_U, +\infty)$ for the constant c_U for which

$$P_0(z > c_U) = \alpha;$$

that is, for $c_U = \Phi^{-1}(1-\alpha)$, or when $\hat{\mu}_2 - \hat{\mu}_1 > \sqrt{m}\sigma\Phi^{-1}(1-\alpha)$. The subscript 0 indicates that the evaluation is under the assumption of the null hypothesis $\Delta\mu = 0$. The condition for sufficient power at $\Delta\mu = \Delta$ is that $P_\Delta(z > c_U) \geq \beta$. The two conditions, for the size and power of the test, yield the inequality

$$m \leq \frac{\Delta^2}{\sigma^2\left\{\Phi^{-1}(1-\alpha) + \Phi^{-1}(\beta)\right\}^2}, \tag{8.1}$$

which is easily solved for the minimum n_1 and n_2, or their total $n = n_1(1 + r)$, obtaining $n_1 = (1 + r)/(rm^*)$ and $n_2 = (1 + r)/m^*$, where m^* is the largest value of m that satisfies (8.1). The sample sizes n_1 and n_2 are then rounded upwards. When some dropout from the study is anticipated, they are inflated so that the sample sizes would be sufficient even after dropout.

While α and β are set by convention, σ^2 has to be guessed and there is usually no obvious protocol for setting the value of Δ. There is plenty of scope for manipulating the process of setting σ^2 and Δ to satisfy a constraint that we may be unwilling to discuss openly or admit that it has an impact on the design, such as limited budget and ability to recruit sufficiently many qualifying subjects. But our principal objections to hypothesis testing are the inflexible management of the rates of error of the two kinds and the intent to act later as if the null hypothesis were valid when it was not rejected.

8.2 Sample Size for Decision Making

We intend to base our inference on the methods introduced in Chaps. 2 and 4, in which we choose the decision that entails smaller expected loss. Apart from the inappropriate decision, we associate costs with experimentation (data collection), and impasse when a range of priors and loss functions is declared.

In a general setting, we could simulate the outcomes of a study for a range of sample sizes (designs) and assess the losses empirically. This is relatively simple to implement, but the computing may take a long time and has to be carefully organised. We choose a computationally more complex but faster method in which we rely on the approximate linearity of the equilibrium function $\delta_0(q)$ in (4.6). Recall that this function describes the priors (δ, q) for which the decision is immaterial in expectation—when the expected losses are identical. Since $\delta_0(q)$ is a function specific to the value of the sample contrast $\widehat{\Delta} = \hat{\mu}_2 - \hat{\mu}_1$, we write $\delta_0(q, \widehat{\Delta})$. We assume that a set of plausible priors (δ, q) is given, and that they form a rectangle, $(q_L, q_U) \times (\delta_L, \delta_U)$.

With $\delta_0(q, \widehat{\Delta})$ a linear function of q,

$$\delta_0(q, \widehat{\Delta}) \doteq \sigma z_R^* \left(\frac{\sqrt{m}}{2} + \frac{q}{\sqrt{m}} \right) - \frac{q}{m} \widehat{\Delta}, \tag{8.2}$$

ignoring the error of approximation (see Sect. 4.2), we can easily establish whether a particular value of $\widehat{\Delta}$ leads to an impasse or not. It suffices to evaluate the balance function at the four vertices of the plausible rectangle. If it has the same sign at all four points, the decision is unequivocal; otherwise we have an impasse. The intercept in (8.2), $s = \frac{1}{2}\sigma z_R^* \sqrt{m}$, does not depend on $\widehat{\Delta}$, so all the equilibrium lines fan out from this single point s at $q = 0$. When s is within the range (δ_L, δ_U), it suffices to check the vertices (q_L, δ_L) and (q_L, δ_U), denoted by LL and LU, respectively. If the balance function has different signs at LL and LU, we have an impasse. This happens for $\widehat{\Delta}$ in an interval $(\widehat{\Delta}_U, \widehat{\Delta}_L)$, the limits of which are easily established by solving the equations $\delta_0(q_L, \widehat{\Delta}) = \delta_S$ for $S = L$ and U. See the left-hand panel of Fig. 8.1 for an illustration; the dashed equilibrium lines correspond to impasse. The solutions are

$$\widehat{\Delta}_S = m\sigma z_R^* \left(\frac{\sqrt{m}}{2q_L} + \frac{1}{\sqrt{m}} \right) - \frac{m\delta_S}{q_L}.$$

In this case $\widehat{\Delta}_L - \widehat{\Delta}_U = m(\delta_U - \delta_L)/q_L$ depends on neither R nor σ. Note that $\widehat{\Delta}_L > \widehat{\Delta}_U$.

When the common intercept s is greater than δ_U, then $\widehat{\Delta}_U$ is the solution of the equation $\delta_0(q_U, \widehat{\Delta}) = \delta_U$ and $\widehat{\Delta}_L$ the solution of $\delta_0(q_L, \widehat{\Delta}) = \delta_L$; see the right-hand panel of Fig. 8.1. When $s < \delta_L$, the vertices LU and UL are in the respective roles of LL and UU. In these scenarios, $\widehat{\Delta}_L - \widehat{\Delta}_U$ depends on both σ and R.

The expectation of the piecewise quadratic loss at an equilibrium prior is

$$Q^* = S^2 \left\{ \left(1 + z_R^{*2} \right) \Phi \left(z_R^* \right) + z_R^* \phi \left(z_R^* \right) \right\}, \tag{8.3}$$

where $S^2 = mq\sigma^2/(m+q)$. For other (plausible) priors, the expected loss is smaller. The expectation of the piecewise linear loss at an equilibrium prior is

$$Q^* = S \left\{ z_R^* \Phi \left(z_R^* \right) + \phi \left(z_R^* \right) \right\},$$

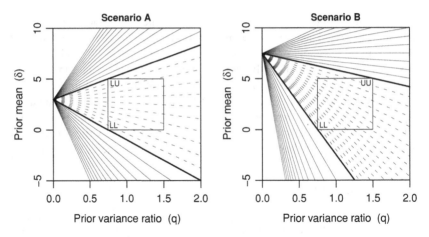

Fig. 8.1 Equilibrium lines that lead to an unequivocal decision (*solid lines*) and to an impasse (*dashes*). Scenarios with the common intercept $s = \frac{1}{2} z_R^* \sigma \sqrt{m} \in (\delta_L, \delta_U)$ in the *left-hand* and $s > \delta_U$ in the *right-hand panel*. The symbol ST marks the vertex (q_S, δ_T) for S, T = L or U

and for the piecewise absolute loss we have the identity $Q^* = \Phi(z_R^*)$.

We adopt Q^* as a criterion for design—we should aim to reduce it below a set threshold or find a means of trading its (low) value for (larger) sample sizes, and the cost of experimentation in general. Another factor to take into account is the probability of impasse; it is reduced with greater sample size, but also by narrowing down the set of plausible priors.

The probability of impasse, related to a plausible prior distribution, is evaluated by integrating the prior distribution of $\widehat{\Delta}$, $\mathcal{N}\{\delta, (m + q)\sigma^2\}$, over the impasse interval $(\widehat{\Delta}_U, \widehat{\Delta}_L)$. We could evaluate these probabilities on a fine grid of values that cover the plausible range, and find their maximum. When $\frac{1}{2}(\widehat{\Delta}_L + \widehat{\Delta}_U) \in (\delta_L, \delta_U)$, the maximum is attained for $\delta = \frac{1}{2}(\widehat{\Delta}_L + \widehat{\Delta}_U)$ and $q = q_L$. When $\frac{1}{2}(\widehat{\Delta}_L + \widehat{\Delta}_U) \notin (\delta_L, \delta_U)$, we evaluate the probability of impasse on a grid of plausible values of q with δ set to either δ_L or δ_U, whichever is closer to the interval $(\widehat{\Delta}_U, \widehat{\Delta}_L)$, or is contained in it.

Example

Suppose elicitation concludes with the plausible priors described by the rectangle $(0.75, 1.5) \times (0, 5)$ for (q, δ) and plausible penalty ratio $(10, 20)$ with piecewise quadratic loss. We assume that $\sigma^2 = 10$ and that $r = n_2/n_1 = 1.5$.

We assemble first all the items required for determining $\widehat{\Delta}_L$ and $\widehat{\Delta}_U$ as functions of $\widehat{\Delta}$. First, $m = 1/n_1 + 1/n_2 = (1 + r)/(n_1 r)$. For $n_1 = 20$, $m = 0.0833$. Next, $z_{10}^* = 0.716$ and $z_{20}^* = 0.926$. The linear approximation to $\sqrt{q(m + q)}$ is quite precise for all plausible values of q; for $q = q_L$, $\sqrt{q(m + q)} = 0.7906$, whereas the

Table 8.1 Sample size calculation for comparing two random samples with sample sizes n_1 and $1.5n_1$, plausible range of penalty ratios (10, 20), $\sigma^2 = 10$ and normal prior distributions for δ with plausible expectation in (0, 5) and $q \in (0.75, 1.5)$; piecewise quadratic loss

	n_1 $(n_2 = 1.5n_1)$									
	10	20	30	40	50	60	70	80	90	100
$R = 10$										
Intercept (s)	0.46	0.33	0.27	0.23	0.21	0.19	0.17	0.16	0.15	0.15
Δ_U	−0.08	0.13	0.18	0.20	0.20	0.20	0.20	0.19	0.19	0.18
Δ_L	1.03	0.69	0.55	0.47	0.42	0.38	0.35	0.33	0.31	0.30
% impasse	14.56	7.67	5.20	3.94	3.17	2.65	2.28	2.00	1.78	1.60
Q^*	2.06	1.09	0.74	0.56	0.45	0.37	0.32	0.28	0.25	0.23
Total cost	5.61	4.11	3.76	3.73	3.83	3.99	4.20	4.42	4.66	4.92
$R = 20$										
Intercept (s)	0.60	0.42	0.35	0.30	0.27	0.24	0.23	0.21	0.20	0.19
Δ_U	0.22	0.34	0.35	0.34	0.32	0.31	0.30	0.29	0.28	0.27
Δ_L	1.33	0.89	0.72	0.61	0.55	0.50	0.46	0.43	0.40	0.38
% impasse	14.56	7.67	5.20	3.94	3.17	2.65	2.28	2.00	1.78	1.60
Q^*	2.65	1.40	0.95	0.72	0.58	0.48	0.42	0.36	0.32	0.29
Total cost	6.20	4.42	3.97	3.89	3.96	4.10	4.29	4.50	4.74	4.98

linear approximation yields $q + \frac{1}{2}m = 0.7917$. Since the factor S is an increasing function of q, we substitute in the expected loss in (8.3) its value with $q = q_U$.

Table 8.1 presents the results for sample sizes $n_1 = 10, 20, \ldots, 100$ and penalty ratios $R = 10$ and 20. The common intercept is listed in the first line to indicate that the scenario A in Fig. 8.1 applies throughout. The ranges of impasse, $\widehat{\Delta}_L - \widehat{\Delta}_U$, therefore do not depend on R and are constant within sample sizes n_1 (columns). The percentage of impasse decreases with n_1, with a steep gradient for small n_1. It is constant within values of n_1 because $\frac{1}{2}(\widehat{\Delta}_L + \widehat{\Delta}_U)$ is a plausible value of δ.

We can decide about the sample size n_1 based on the (conservatively assessed) expected loss in tandem with the prior-related percentage of impasse. A more profound approach takes into account the cost of experimentation and quantifies on the same scale the value of small Q^* and the threat of impasse. By way of an example, suppose impasse is associated 12 units of loss, expressed in the same units as used for the expected loss Q^*, and the cost of the experiment is $1.5 + 0.012n_1(1 + r)$, that is, 1.5 units for the general setup and 0.012 for each unit used. For instance, for $R = 10$ and $n_1 = 10$, the overall cost is $1.5 + 12 \times 0.1456 + 2.06 + 0.30 \doteq 5.61$. The corresponding values for the other settings (R, n_1) are given in the bottom line of each block of Table 8.1. The minimum total cost is attained for n_1 around 40, that is, for $n = n_1 + n_2 \doteq 100$.

We do not have to be satisfied with this proposal, especially if it is based on a value of σ^2 for which we have pretended certainty. It may also be prudent to re-calculate the total cost with alternative values of the factors applied in Table 8.1. We should not rush into a decision based on the settings which we have pretended to be ideal. For example, by calculating the quantities in Table 8.1 for $\sigma^2 = 12$ we find that the lowest total cost is still attained for $n_1 \doteq 40$, although a more detailed evaluation would find that it has increased by a unit, from 37 to 38 for $R = 10$ and from 40 to 41 for $R = 20$.

By a similar approach we can explore how much we could save by having a narrower range of plausible prior parameters, for the prior mean δ in particular. Suppose the plausible range of δ is $(2, 4)$ and the range of q is unchanged. Then the total cost is minimised for $n_1 = 27$ with $R = 10$ and 30 with $R = 20$. The total cost and its components are drawn in Fig. 8.2 as functions of the sample size n_1. The total costs based on $\sigma^2 = 10$ and $\sigma^2 = 12$, both drawn by gray colour, with a thinner line for $\sigma^2 = 12$, are very close to each other; the total cost with the narrow prior (and $\sigma^2 = 10$) is much lower. Note that the expected loss (due to bad decision) is very small (not distinguishable in the diagram for the wide and narrow range of plausible priors) in relation to the cost of the impasse. Of course, the cost of experimentation is the same function of n_1 for the two sets of plausible priors.

The total cost as a function of n_1 is very flat at its minimum, so we do not have to insist on the exact number of units; the consequences of a small deviation in the design in either direction are trivial. The sample size required for $R = 20$ is only slightly greater than for $R = 10$. However, the assumptions about the prior make a lot of difference, and should therefore be made with care and integrity. The minima of the functions for the three settings are marked by vertical ticks on the function and at zero height.

Since all the calculations in Table 8.1 are in scenario A (Fig. 8.1), reducing the upper bound q_U is not important, but increasing q_L might appear useful. However, with a greater value of q, we should revise (and reduce) the value of σ^2. If we do so, then the increase of q_L conveys no advantage.

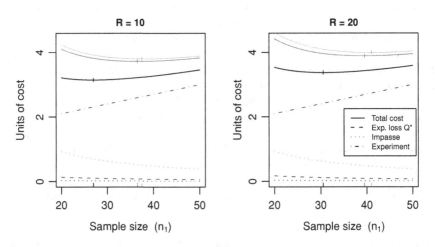

Fig. 8.2 The total cost of experimentation as a function of the sample size with a wide range of plausible priors, $\delta \in (0, 5)$, drawn in *gray* colour, thinner for the version based on $\sigma^2 = 12$, and a narrow range for δ, $(2, 4)$, drawn in *black*. The other parameters are the same as in Table 8.1

The procedure for piecewise linear or constant loss differs only by setting z_R^* to the solution of the balance equation (4.7) instead of (4.5). For the piecewise constant loss, we have $z_R^* = \Phi^{-1}\{R/(R+1)\}$.

With a set design, we can simulate the planned study for a range of plausible values of $\Delta\mu$ to get a feel for the likely results. We may assess the robustness of our procedure by using settings that deviate from those used in sample size calculation. For example, dropout in a study in which human subjects are used is often a well-founded concern. Thus, instead of the sample size of $n = 100$, we set it to 125, anticipating 20 % dropout. The subjects are randomised to groups in proportion 2:3 (50+75), but then the simulated dropout in the respective groups is 25 and 16 %. The number of subjects lost is not constant in replications, but binomial with parameters (50, 0.25) and (75, 0.16), The respective standard deviations are 3.1 and 3.2.

Such a simulation is useful for introspection, to reassess whether the range of loss functions and the criterion for the design have been defined appropriately. Also, some insight can be gained into what may happen when something goes wrong. Of course, we have to rely on our own imagination to think of the various contingencies in advance. Invaluable in this enterprise is the collection of accounts from similar experiments conducted in the past, and the associated 'detective' work in general. Such a practice could be promoted by recording the details not only of successes but also of failures in the study design and execution, and by enhancing the research ethos in which failures cause no embarrassment, so long as they are well recorded, lessons from them learned and, perhaps, the failures not repeated.

In a simulation, we record only the decision: A or B (unequivocal) or impasse. Figure 8.3 summarises sets of 10,000 replications of such decisions based on values of δ in the range $(-1, 1)$ for $R = 10$ and $R = 20$. The shaded part represents impasse. The cross-hairs at $(0, 0)$ are added to make the two panels easier to compare. One

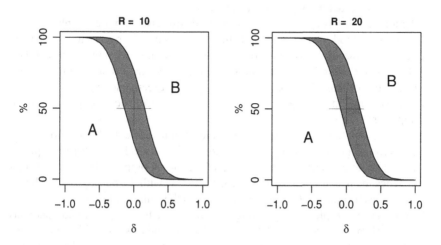

Fig. 8.3 The probabilities of the decisions A, B and of impasse (*shaded area*) for plausible priors $q \in (0.75, 1.5)$ and $\delta \in (0.5)$

might be concerned that the probability of impasse is rather large for some values of δ. (The width of the gray area should be measured vertically.) For example, it exceeds 50 % around $\delta = 0$ for $R = 10$ and for $\delta \in (0, 0.1)$ for $R = 20$. The design can accommodate this concern, or second thought, by raising the cost of impasse.

At the design stage, we can vaccilate about the costs and losses and rerun the sample size calculations many times, but ultimately we have to settle for plausible ranges of all the parameters on which the sample sizes are based. To produce Fig. 8.3 takes about 3 min of CPU time, but a coarser version of the diagram, using 21 values of δ instead of 41 and 2,000 replications instead of 10,000, can be produced in about one tenth of the time. Thus, repeating the simulations with a range of settings is feasible, and is very useful for studies with expensive data collection. Once the data collection stage of the study has commenced, the settings are fixed and we are fully committed to them. Regret may be avoided by constructive pessimism and exhaustive exploration of plausible scenarios at the planning stage.

8.3 Problems, Exercises and Suggested Reading

1. Derive the optimal sample sizes for testing a two-sided (symmetric) hypothesis of equal expectation of two normal samples with equal variance. Explore how this method can be extended to samples from other distributions, such as the Poisson and beta, in which the expectation and variance are related.

2. Check the results of sample size calculations in Sect. 8.1 by simulations. That is, generate a large number of replicate datasets with the calculated sample sizes and check that the planned test has the specified power. This exercise can be made more realistic by adding some twists that occur in practice, such as nonresponse and some (moderate) heteroscedasticity.

   A more ambitious task (project): Repeat this exercise with the sample size calculation method in Sect. 8.2, evaluating the minimum-expected loss decisions. Set the sample size by one criterion, but assess the simulation results by another.

3. Describe the difficulties that would arise in sample size calculation in Sect. 8.2 if the set of plausible prior parameters was not a rectangle; e.g., if it were the inside of an ellipse or a general polyhedron. Why would convexity of this set be important?

4. The loss function has a single role in the sample size calculation—to set z_R^*. Therefore, a given value of z corresponds to different values of R for the piecewise quadratic, linear and constant loss functions. Compile a function in R that matches the penalty ratio R for the quadratic loss with the penalty ratio in linear (and constant) loss so that they correspond to the same value of z^*.

5. Consider the problem of designing a study in which two or more comparisons of sample means are planned. Compare the problem of setting the sample sizes for hypothesis testing (adequate power for a given test size) and for making decisions. Address this problem first for independent contrasts (e.g., comparing

group 1 with 2 and group 3 with 4), and then for dependent contrasts (e.g., comparing group 1 with 2 and 1 with 3).

6. Describe the difficulties with sample size calculation (for both methods in Sects. 8.1 and 8.2) when the variance σ^2 is not known and will be estimated. Discuss the compromise made in the calculation for hypothesis testing, and whether (and how) it can be adapted in the calculation with loss functions.

7. Suggested reading: Lenth (2001) and Julious (2009). Unsurpassed classics on study design: Kish (1965) and (1987).

8. A project: Adapt the method of sample size calculation to binomial (or Poisson distributed) outcomes.

9. A project: Collect information about the costs of the the incorrect decision in an experiment that you regard as useful to conduct. Record all the elements of the cost, together with the uncertainties involved. Consult a suitable authority. Record also the expenses on the experiment and the losses due to an impasse.

10. Suggested reading about design for small-area estimation: Longford (2006) and (2013).

References

Julious, S. A. (2009). *Sample sizes for clinical trials*. London, UK: Chapman and Hall/CRC.

Kish, L. (1965). *Survey sampling*. New York: Wiley.

Kish, L. (1987). *Statistical design for research*. New York: Wiley.

Lenth, R. (2001). Some practical guidelines for effective sample size determination. *The American Statistician, 55*, 187–193.

Longford, N. T. (2006). Sample size calculation for small-area estimation. *Survey Methodology, 32*, 87–96.

Longford, N. T. (2013). Sample size calculation for comparing two normal random samples using equilibrium priors. Communications in Statistics - Simulation and Computation 42 (to appear).

Index

N. T. Longford, *Statistical Decision Theory*,
SpringerBriefs in Statistics, DOI: 10.1007/978-3-642-40433-7,
© The Author(s) 2013